Almost All About Waves

John R. Pierce

Dover Publications, Inc.
Mineola, New York

Bibliographical Note

This Dover edition, first published in 2006 by special arrangement with
the MIT Press, is an unabridged republication of the 1981 second printing of
the work originally published in 1974 by The Massachusetts Institute of
Technology, Cambridge, Massachusetts.
Figure 3.1 on page 19 is reproduced from *Electromagnetic Waves* by S. A.
Schelkunoff (New York: D. Van Nostrand Company, Inc., 1943).
Figure 4.9 on page 33 is reproduced from *The Speech Chain* by P. B. Denes
and E. H. Pinson (Bell Telephone Laboratories, 1963).

International Standard Book Number

ISBN-13: 978-0-486-45302-6
ISBN-10: 0-486-45302-2

www.doverpublications.com

To Ed and Ann David,
who have been good to me
and good for me.

Foreword

Today physicists and engineers have at their disposal two great tools: the computer and mathematics. By using the computer, a person who knows the physical laws governing the behavior of a particular device or a system can calculate the behavior of that device or system in particular cases even if he knows only a very little mathematics. Today the novice can obtain numerical results that lay beyond the reach of the most skilled mathematician in the days before the computer. What are we to say of the value of mathematics in today's world?

Mathematics is still as fascinating as it ever was, and is worthy of pursuit for those whom it fascinates. But what of the person with a practical interest, the person who wants to *use* mathematics?

Today the user of mathematics, the physicist or the engineer, need know very little mathematics in order to get particular numerical answers. Perhaps he can even dispense with the complicated sort of functions that have been used in connection with configurations of matter. But, a very little mathematics can give the physicist or engineer something that is harder to come by through the use of the computer. That thing is *insight*.

The laws of conservation of mechanical energy and momentum can be simply derived from Newton's laws of motion. The laws are simple; their application is universal. There is no need for computers, which can be reserved for more particular problems.

Similarly, we can learn a lot about waves through very simple mathematics, as I hope that I demonstrate in this book. The book does not tell you how best to get the answer to a particular complicated problem. It tells you what *sort* of answer the answer must be, for example, what must happen when waves are coupled, and what relations must hold among the energy and momentum of waves.

A few years ago, Harald T. Friis provided a very simple derivation for his radio transmission formula, something that clever mathematicians and physicists had failed to do. This led me to believe that many other facts about waves could be derived with very little mathematics. The fascination of the challenge led me to write this book.

The book was not written without a considerable struggle to understand certain matters. The writer is particularly indebted to Professor Hermann A. Haus and Professor Charles H. Papas for very substantial help, and especially to Dr. Tse Chin Mo, who read, commented on, and made valuable suggestions concerning two versions of the text.

1 The Idea of Waves

The idea of waves is one of the great unifying concepts of physics. Men must have observed waves from the earliest times. In the fifteenth century, Leonardo da Vinci wrote of waves, "The impetus is much quicker than the water, for it often happens that the wave flees the place of its creation, while the water does not; like the waves made in a field of grain by the wind, where we see the waves running across the field while the grain remains in place." Clearly, Leonardo recognized that when a wave of water moves from one place to another the water does not go bodily with it.

Modern physics is full of waves: the earthquake waves which seismologists study; the waves and ripples on oceans, lakes, and ponds; the waves of sound which travel through the air; the mechanical waves in stretched strings and in the quartz crystals that are used to control the frequency of radio transmitters; the electromagnetic waves that constitute light, and that are radiated by radio transmitters and received by radio receivers; and finally, the waves of what?—probability, perhaps—which are used in quantum mechanics to predict the behavior of electrons, atoms, and complex substances.

What are waves? They are not earth, or water, or air; steel, or catgut, or quartz; yet they travel in these substances. Nineteenth-century physicists felt constrained to fill the vacuum of space with an ether to transmit electromagnetic waves, yet so arbitrary a substance seems more a placebo to quiet the disturbed mind than a valid explanation of a physical phenomenon. When we come to the waves of quantum mechanics, the physicists do not even offer us a single agreed-upon physical interpretation of the waves with which they deal, although they all agree in the way they use them to predict correctly the outcome of experiments.

Rather than asking what waves are, we should ask, what can one say about waves? Here there is no confusion. We recognize in waves a sort of behavior that can be described mathematically in common terms, however various may be the physical systems to which the terms are applied. Once we recognize that in a certain phenomenon we are dealing with waves, we can assert and predict a great deal about the phenomenon even though we do not clearly understand the mechanism by which the waves are generated and transmitted. The wave nature of light was understood, and many of its important consequences were worked out, long before the idea of an electromagnetic wave through space was dreamed of. Indeed, when the electromagnetic explanation of the physical nature of light was proposed, many physicists who recognized clearly that light was some sort of wave refused to accept it.

We can study the important principles of waves in simple and familiar examples. As we come to understand the behavior of these waves, we can abstract certain ideas that are valid in connection with all waves, wherever we may find them.

One thing is immediately obvious concerning waves. Like moving objects, moving waves carry energy from one point to another.

The electromagnetic waves of light and heat which come to the earth from the sun have a power of about one kilowatt per square meter. Solar cells can convert about a tenth of the solar energy that falls on them into electrical energy. By means of solar cells, the energy of sunlight powers communication satellites and other space vehicles. Growing plants convert energy of the sun's electromagnetic waves into chemical energy. When we burn wood or coal, we release this energy and make use of it.

Television transmitters send forth electromagnetic waves whose power is tens of thousands of watts. Each of many television receivers picks up a minute fraction of this transmitted power. The waves of the ocean beat on the shore with tremendous energy; during storms the sea moves rocks that weigh tons. The power of the waves of sound that the human voice generates is minute.

In every case, the waves do carry energy from one point to another, however much the amount of energy may differ from case to case.

Like moving matter, moving waves have momentum. When waves are absorbed by or reflected from an object, they push on it. Ordinarily, the momentum of waves is less noticeable than their energy. Yet, in May 1951, Russell Saunders published an article, "Clipper Ships of Space" in *Astounding Science Fiction* which showed that it was theoretically possible to move a spaceship through the solar system by the pressure of light on huge sails. And, physicists must take the momentum of sound waves and light waves into account in understanding the properties and behavior of solid substances.

Finally, it takes time for a wave to travel from one point to another. That is, waves have velocity. Light waves travel very swiftly—186,000 miles a second or 300,000 kilometers a second. The sound waves which travel to our ears through the air move more slowly—about 836 miles per hour or 1129 feet per second. Sound travels through water and solids at a higher speed. Waves on the surface of the water travel more slowly.

Energy, momentum, and velocity are important properties of waves. Another and truly astounding property that many waves exhibit is *linearity*.

When you throw two pebbles into a pond, the expanding circles of ripples do not affect one another. One set of ripples passes right through the other. While the intersecting pattern may look complicated to the eye, it can be seen as two independent sets of expanding circles. When two people speak to one another, the sound waves of their voices do not rebound from one another; they pass through one another. The faint rays of the stars are not affected by the bright sunshine they traverse on the way to our night sky.

Waves that do not affect the passage of other waves are called *linear* waves because the total of two waves is simply the sum of the waves as they exist separately. As two linear waves with crests of heights H_1 and H_2 pass, the greatest height is simply $H_1 + H_2$—a linear relation. For linear waves, a wave of height $2H$ is the same as two waves of height H existing in the same place at the same time. For a linear wave, the velocity cannot depend on the height or strength of the wave, for a large-amplitude wave can be regarded as simply the sum of a number of small-amplitude waves.

Few waves are exactly linear. Many waves (the waves of sound, for example) are so nearly linear over the common range of intensity that we get almost perfectly exact answers by treating them as if they were linear. In this book we shall deal almost exclusively with linear waves.

We are all familiar with echoes. When we shout toward a vertical surface, either the face of a building or the face of a cliff, the sound of our voice is reflected by the solid surface and returns to us. This is much like a ball bouncing from a wall. There are differences, however.

One difference we have just discussed. If we threw many balls at a wall simultaneously, the balls might be "reflected" from (bounce off) one another as well as from the wall. Linear waves (and sound waves of moderate loudness are linear) pass through one another. Thus, if we shout in a large, hard-walled room we hear endless echoes of the sound from the walls, echoes that finally die away into a weak, confused murmur. Acousticians speak of the *reverberation time*, which is the time in which the reverberation dies down to a millionth of its initial power or energy.

The reflection of waves differs in another way from the "reflection" of solid objects such as bouncing balls. A solid object has a permanent form which it retains after reflection. What about a wave?

The shape of a wave can be greatly altered when it is reflected. If a wave strikes a rough surface it can be scattered or reflected in many directions. A rough surface may reflect a ball in an unpredictable direction, but not in all directions at once. We shall not try to unravel all the complexities of the reflection of waves. However, one special case is at once very important and very simple.

Let us represent a wave that approaches a surface by a curve such as that shown in Figure 1.1. This can represent how pressure, or electric field, or the displacement of a stretched string varies with distance in the z direction. Let us assume that this whole pattern travels to the right (in the $+z$ direction) with a velocity v, and that it hits a perfectly reflecting surface. What does the reflected wave look like?

Two simple possibilities are shown in Figure 1.2. In each case the wave travels from right to left and the shape of the wave has been turned left to right. This is reasonable, because the part of the wave which hits the reflector first must come back first.

Beyond this, the wave may be reflected as in a or as in b. The reflection shown in b is the negative of the reflection shown in a. If the original wave had a higher than average pressure, the reflected wave of b has a lower than average pressure. Similarly, on reflection a positive field can be changed into a negative

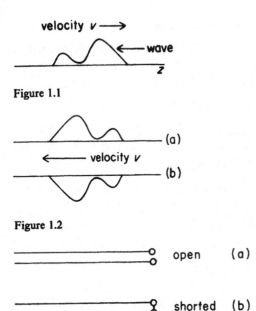

Figure 1.1

Figure 1.2

Figure 1.3

field, or an upward displacement of a string into a downward displacement of a string. Is this reasonable?

Let us think of the wave as a voltage that travels down a transmission line. If the line is open at the end, as shown in a of Figure 1.3, the wave is reflected, and the reflection is as shown in a of Figure 1.2. However, if the line is shorted at the end, as in b of Figure 1.3, the wave is reflected as in b of Figure 1.2. Why is this?

If the line is shorted at the point of reflection, the voltage must be zero at that point. Hence, during the process of reflection the voltage of the incident wave plus the voltage of the reflected wave must always be zero. Hence, at every instant during the process of reflection, the reflected wave must be the negative of the incident wave.

Consider a wave that is the sidewise displacement of a

stretched string. If such a wave is reflected from a solid object to which the string is tied, the reflected wave must be the negative of the incident wave, as in b of Figure 1.2.

When a guitar string is plucked, a wave is generated which runs back and forth between the ends of the string. The wave is reflected repeatedly at each end. At each end, the direction of travel is reversed and the reflected wave is made the negative of the incident wave. When the wave has been reflected twice, once at each end of the string, it has its original direction and shape. If the length of the string is L, the wave generated by plucking the string must travel a distance $2L$ in order to get back to its initial position and direction of travel and shape after reflection at each end. If v is the velocity of the wave, the wave traverses this distance $(2L)$ f times per second, where

$$f = \frac{v}{2L}.\tag{1.1}$$

The frequency f is the pitch of the string. If we put a finger down on a fret that halves the length of the string, the pitch is doubled and the note goes up an octave, that is, the frequency is doubled.

The ratio of the pitches of two semitones is the twelfth root of two. This determines the spacing of the frets of a guitar. The ratio of the distance L_1 and L_2 of successive frets from the bridge or soundbox end of the guitar string is always

$$\frac{L_2}{L_1} = 2^{1/12} = 1.05946.\tag{1.2}$$

Figure 1.4 illustrates this.

The velocity with which a wave travels along a stretched string, as in a guitar or a piano, depends on the tension of the string and on the mass per unit length of the string. The higher

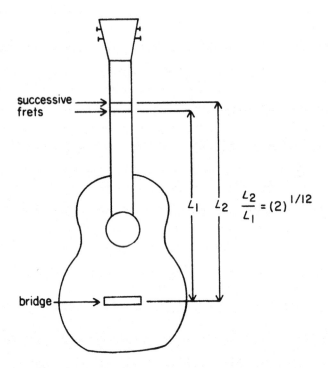

successive frets

bridge

$$L_1 \quad L_2 \quad \frac{L_2}{L_1} = (2)^{1/12}$$

Figure 1.4

the tension, the higher the velocity and the higher the pitch. The more massive the string, the lower the velocity and the lower the pitch. The strings of the low notes on the piano are wound with wire in order to make them more massive. This lowers the velocity and pitch.

In this chapter we have discussed a number of things concerning waves and have considered special cases of reflection in some detail. In later chapters we shall consider further some of the matters that have been raised in this chapter. We shall find that there are two sorts of velocity: phase velocity and group velocity. We shall find that we must consider the momentum of waves very carefully in order to understand it. We shall find that it is easy to define the energy, momentum, and velocity only in terms of a particular, simple sort of wave, a sinusoidal wave. Fortunately, other sorts of waves can be represented as sums of sinusoidal waves.

Going on, we shall see some of the consequences of those properties of waves we have come to understand.

Microwave radio and radar systems make use of *directional couplers* which transfer part of a wave from one waveguide to another. We shall find that we can understand the performance of such devices without considering the particular ways in which electromagnetic waves differ from other waves.

Tweeters and some other speakers make use of horns. We shall learn important things about the frequency response of horns.

Communication satellites and many other microwave communication systems make use of *traveling wave tubes* to amplify the signals they transmit. In a traveling wave tube, the microwave output energy is derived from a beam of moving electrons. We shall find that we can understand the mechanism of amplification by studying the energy and momentum of waves and the energy of waves on a moving medium; we do not need to consider details peculiar to waves on an electron beam.

Parametric amplifiers are often used to amplify very weak microwave signals. In a parametric amplifier, the source of gain is a parameter (inductance, capacitance, dielectric constant, or wave speed) which changes with time. We shall be able to understand in very simple terms the circumstances under which a strong, nonlinear wave can push on weak linear waves and cause them to grow with distance.

We shall be able to understand how waves can be radiated in various directional patterns, as they are by radio antennas, and we shall find the fraction of the transmitted power that is received by a microwave antenna.

We shall learn something about the power radiated by objects which move faster than the speed of a plane wave—whether these objects be electrons traveling through a dielectric or boats traveling on water.

Thus, in the remaining chapters of this book we shall consider both fundamental properties of waves, and the practical effects of these properties in a wide variety of applications.

Problems

1. Assume that a wave has a simple rectangular shape instead of the smooth shape shown in Figure 1.1. Draw a picture representing the wave in successive stages of reflection assuming (a) reflection according to a of Figure 1.2 and (b) reflection according to b of Figure 1.2. Some of the plots should show an overlapping of the incident and reflected waves.

2. For an acoustic wave in a tube such as an organ pipe, the reflection of the sound pressure at the end is as a of Figure 1.2 when the tube is closed at the end and according to b of Figure 1.2 when the pipe is open at the end. Find an expression for the pitch in terms of length of the tube and velocity of the wave for the following cases: (a) tube open at both ends; (b) tube open at one end and closed at the other end; and (c) tube closed at both ends.

3. An organ pipe is open at one end and closed at the other end. How long should the pipe be to produce a pitch of 440 hertz (A above middle C)?

4. What is the pitch of (a) a 10-foot pipe closed at both ends, (b) a 10-foot transmission line that is shorted at both ends?

5. Sometimes a device is used to clamp the strings of a guitar between frets. What will happen if the same fingering relative to the clamp is used below the clamp as was used below the top end of the strings without the clamp?

6. A tower 500 feet high acts as a radio antenna. The ground at the bottom is a good conductor. What is the "pitch" (frequency at which the antenna is resonant)?

7. The pitch ratio in going from one key to the next (white *or* black) on a piano is $2^{1/12}$ or approximately 1.05946. A piano (equally tempered or well tempered) fifth is 7 half steps; a piano fourth is 5 half steps; a piano third is 4 half steps. The frequency ratio for an ideal fifth is 3/2; for an ideal fourth is 4/3; for an ideal third is 5/4. What is the percentage of errors in the piano, or tempered, fifth, fourth, and third?

2 Sinusoidal Waves—Their Strength and Power

When we look at the stormy sea, we realize that waves can be very complicated indeed. In understanding things it is always best to consider simple examples first. Hence, we shall first consider *guided* waves, such as the mechanical waves that travel along a stretched string, or the electromagnetic waves that travel through coaxial cables, rather than waves on the surface of the sea, or sound waves or radio waves which travel in an open space.

Further, we shall first consider waves that vary in a particular, simple way in space and time as they travel along the guiding medium. The waves that we shall consider are sinusoidal waves.

In discussing such waves we must make use of a quantity that specifies the strength of a wave at a particular time and place. This strength may be the strength of an electric field or of a magnetic field. It may be the voltage or current at a given time and place. It may be the pressure or the velocity associated with a sound wave. It may be the transverse displacement of a stretched string or the rate at which that displacement changes with time.

Whatever the nature of the wave, and whatever the measure of the strength of the wave at a particular time and place, we shall use the same symbol, S, to designate this strength.

Let us, then, proceed to the discussion of *any* sinusoidal wave that travels in a direction which we shall call the z direction. It is convenient to use the cosine function rather than the sine function, so the strength S of our sinusoidal wave varies with time and distance as

$$S = S_0 \cos(\omega t - kz + \phi). \tag{2.1}$$

Here S_0 is a constant that is equal to the *peak strength* (or *peak amplitude*) of the wave; t is time and z is distance in the direction in which the wave travels; ω is called the *radian frequency* and k is called the *wave vector* or the *phase constant*; ϕ is a constant giving the *phase* of the wave.

We see that we can express S in terms of an angle θ

$$S = S_0 \cos \theta, \tag{2.2}$$

where θ is given by

$$\theta = \omega t - kz + \phi. \tag{2.3}$$

As either t or z changes, θ changes in accord with (2.3) and S changes in accord with (2.2). The variation of S with θ is shown in Figure 2.1; this is just a plot of $S_0 \cos \theta$ versus θ.

From (2.2) we notice that S is constant if $\omega t - kz$ is constant. That is, S is constant if

$$z = vt, \tag{2.4}$$

where

$$v = \omega/k; \tag{2.5}$$

v is called the *phase velocity* of the wave. As the wave moves, the whole sinusoidal pattern moves to the right with a velocity v. If we move to the right with a velocity v and observe the wave, we see the sinusoidal pattern of the wave as a stationary ripple.

If we stand still and watch the wave, we observe a sinusoidal displacement traveling to the right. Figure 2.2 shows the variation of S with distance at the time $t = 0$.

The distance between wave crests is the *wavelength* λ. If the velocity with which the wave travels is v, then frequency f with which crests pass us is

$$f = \frac{v}{\lambda}. \tag{2.6}$$

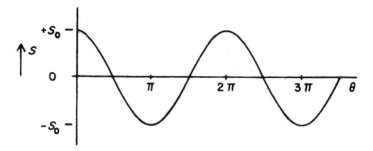

Figure 2.1

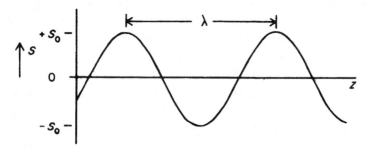

Figure 2.2

The radian frequency ω is 2π times f, and f crests pass a given point each second. This means that the angle θ of (2.3) must change $2\pi f$ radians each second. Thus

$$\omega = 2\pi f. \tag{2.7}$$

If we move from one crest of the wave to the next, the distance we move with respect to the wave is one wavelength, that is, a distance of λ. The angle ω must change by 2π radians in moving from crest to crest, so the following relation must hold:

$$k\lambda = 2\pi,$$

$$k = \frac{2\pi}{\lambda}. \tag{2.8}$$

The frequency f is the number of crests which pass a point in one second. Clearly, the time is called the *period* T of the wave:

$$T = 1/f. \tag{2.9}$$

From (2.6) and (2.9) we see that

$$v = \frac{\lambda}{T}. \tag{2.10}$$

Here we have velocity expressed as the distance λ between crests divided by the time T between the passage of successive crests.

The foregoing relations are concerned with the form and behavior of a wave as described in terms of its strength S. *Power* is another important aspect of waves. In generating a radio wave we must continually supply so many watts or kilowatts of power which the wave carries away. The wind supplies power to the waves it produces in the sea. Electric power must be supplied to a speaker in order to generate sound waves.

The power of a wave is proportional to the square of its strength. Consider the wave described by Equation 2.1

$$S = S_0 \cos(\omega t - kz + \phi). \tag{2.11}$$

The power at any instant of time, P_i, is proportional to S^2. We can choose the units in which we measure S so that

$$P_i = 2S^2 = 2S_0^2 \cos^2(\omega t - kz + \phi). \tag{2.12}$$

By using a simple trigonometric relation,

$$\cos^2\theta = \tfrac{1}{2}(1 + \cos 2\theta),$$

we obtain from (2.12)

$$P_i = S_0^2[1 + \cos 2(\omega t - kz + \phi)]. \tag{2.13}$$

We see from (2.13) that the instantaneous power of a wave

varies with time. The variation has a radian frequency 2ω. Power is greatest when the strength, as given by (2.11), has its extreme positive *or* negative value, $+S_0$ or $-S_0$, because

$$(+S_0)^2 = S_0^2 = (-S_0)^2.$$

In (2.13) the average value of $\cos 2(\omega t - kz + \phi)$ is zero. Hence, the average power, which we shall call P, is simply

$$P = S_0^2. \tag{2.14}$$

In later work we shall frequently call the average power P the power and be unconcerned with the fact that the power actually fluctuates between 0 and $2P$ with a frequency 2ω. We should, however, keep in mind that the power of a wave does vary periodically, just as the strength does, but with twice the frequency with which the strength varies.

It may be convenient to tabulate some of the relationships of this chapter. Concerning radian frequency ω, frequency f, wave vector k, phase velocity v, wavelength λ, and period T we can say

$$S = S_0 \cos \theta,$$

$$\theta = \omega t - kz + \phi,$$

$$v = \omega/k = \lambda/T = \lambda f,$$

$$\omega = 2\pi f,$$

$$k = 2\pi/\lambda.$$

Concerning instantaneous power P_i and average power P we can say

$$P_i = S_0^2(1 + \cos 2\theta),$$

$P = S_0^2$.

These relations concerning power hold only if S_0 is measured in units such that

$P_i = 2S^2$.

Power is a quantity which has direction. A power P in the $+z$ direction is the same as a power $-P$ in the $-z$ direction.

Problems

1. Starting with $S = S_0\sin(\omega t - kz + \theta)$ derive the relations corresponding to (2.12), (2.13), and (2.14).

2. Assume that two waves are present, so that

$$S = S_1\cos(\omega_1 t - k_1 z + \theta_1) + S_2\cos(\omega_2 t - k_2 z + \theta_2).$$

If $P_i = 2S^2$, what are the instantaneous and average powers, assuming that $\omega_1 \neq \omega_2$.

3. In which direction does a wave $S = S_0\cos(\omega t + kz + \theta_1)$ travel? What are the instantaneous and average powers?

4. What are the instantaneous and average powers of the wave

$$S = S_0[\cos(\omega t + kz) + \cos(\omega t - kz)]\ ?$$

3 Media and Modes

A stretched string is a *medium* along or through which waves can travel. When we think of waves in such a string, we usually think of the transverse waves we generate by plucking or bowing a string. Other waves can travel through a string.

If we hit a long bar on one end (parallel to the axis), the bar rings because *longitudinal* waves travel back and forth along the bar. If we suddenly twist the end of a long bar, *torsional* waves travel along it.

If we consider electromagnetic waves traveling through a pipe or *waveguide*, we find that many different field patterns can travel through the pipe, each with a different velocity. Figure 3.1 shows some field patterns that can exist in tubes or pipes.

If, in considering waves, we focus our attention on the medium that supports the waves, we encounter all sorts of complexities. Happily, there is another approach to the study of waves. Much of the behavior of waves doesn't depend on the particular nature or pattern of the disturbance that constitutes the wave. We can learn much about waves by characterizing the wave simply as a particular sort of disturbance, or *mode* of a medium through which waves can travel. Indeed, the equations of the preceding chapter describe one particular wave or mode among many that might travel in a medium.

In considering waves on a stretched string, we can characterize a transverse sinusoidal wave in the vertical plane as one mode and a wave in the horizontal plane as another mode. But, we can go further than this. A transverse sinsoidal wave can travel along a string in one of two directions: it can travel *forward* with a velocity v, or *backward* with a velocity $-v$. We shall regard these forward and backward waves as two distinct modes, even though they differ only in direction of travel.

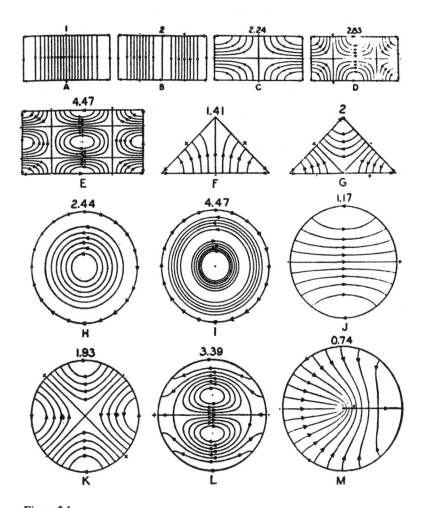

Figure 3.1

We shall, however, regard similar sinusoidal waves of different frequencies as representing the same mode if, when we change the frequency slightly, the wave changes only slightly in form or in velocity.

If we want to express any possible disturbance on or excitation of a medium, we must take into account all possible modes and all possible frequencies. But, many useful phenomena can be understood by considering only one or two modes. It is not surprising that this is simpler than trying to deal with a more general excitation of the medium.

The behavior we consider will be characteristic of many different sorts of waves, such as waves traveling along stretched strings, or sound waves, or electromagnetic waves. When we use the word wave, we shall mean such a mode. As we have noted, the equations of Chapter 2 describe a single mode. So will later equations unless it is explicitly stated that we are considering more than one mode.

Here we shall take mode to mean a wave that travels with constant velocity along a medium that does not change with distance. Mode usually has a somewhat broader meaning. In Chapter 9 we shall consider waves in a tapered medium. These can be thought of as modes. We shall think of them as made up of the modes of an untapered medium.

The reader should be cautioned that the work in this book applies to modes in the narrow sense of waves that travel with a constant velocity in a medium that does not change with distance. Some of the work, but not all, applies to the modes of tapered media as well.

Problems

1. How many modes has a stretched string?

2. A "shielded pair" has two wires in a metal tube. How many modes does this have?

3. A three-phase power transmission system has three parallel wires above ground. How many modes are there?

4. Over a certain range of pitch, pianos are strung with three strings per note. How many modes are there among three such strings? How many different velocities or pitches can there be?

4 Phase Velocity and Group Velocity

In Chapter 2 we considered the velocity v of the wave, which may be expressed in various ways. From (2.5)–(2.10) we see that

$$v = \lambda f = \frac{\lambda}{T} = \frac{\omega}{k}.$$

Sinusoidal waves of different frequencies (or periods) have different velocities v or wave vectors k. The relation between frequency (or period) and wave vector (or wavelength) can be expressed by plotting ω vs k. Such a plot is shown in Figure 4.1. For a given point on the curve, that is, for a given radian frequency ω and wave vector k, the velocity v is equal to ω/k. This is just the slope of a straight line drawn from the origin to the point k,ω.

In some cases the plot of ω vs k is a straight line, as shown in Figure 4.2. In this case the slope of a straight line from any point on the curve to the origin is the same, that is, it is the slope of the straight line that is the plot of ω vs k. When this is so, the velocity v is the same for waves of all frequencies.

Modes for which v is independent of frequency are called *nondispersive* modes. Let us see why such a name was chosen.

In Figure 4.3 we have a plot, for a given instant, of the

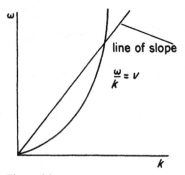

Figure 4.1

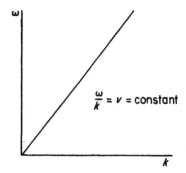

Figure 4.2

strengths of two waves, a and b, which have different frequencies and wavelengths. The sum of these waves at the same instant is the total strength, shown as c.

If the two component waves a and b travel to the right at the same velocity v, their sum c maintains the same shape and simply travels to the right at the velocity v. But, if the velocities of a and b were different, the waves a and b would slide past one another as they traveled, and their sum would change shape as the sinusoidal waves traveled along.

The excitation of a mode can be a short *pulse* as shown in Figure 4.4. Such a pulse can be considered as made up of many sinusoidal component waves having different frequencies. If the mode is nondispersive, that is, if the sinusoidal component

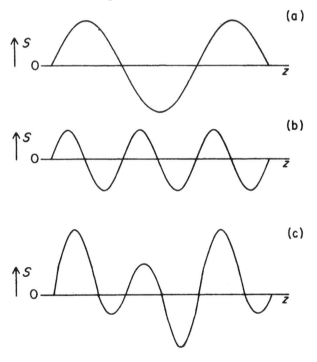

Figure 4.3

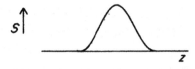

Figure 4.4

(a) low- frequency modulating signal

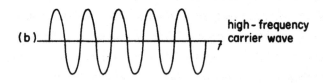

(b) high - frequency carrier wave

(c) amplitude - modulated signal

Figure 4.5

waves of different frequencies travel with the same velocity, all the sinusoidal component waves travel together, and the pulse retains the same shape as it travels along. However, if the velocity v is different for different frequencies, some sinusoidal frequency components lag behind or run ahead of others, and the pulse is broadened or dispersed as it travels. Hence, a *dispersive* mode is one for which the velocity changes with frequency, so that pulses are broadened or dispersed as they travel. Conversely, if the velocity v does not change with frequency (that is, if the ω vs k curve is a straight line through the origin), the mode is nondispersive and the pulse retains its original compact form as it travels.

Some complicated signals are made up of frequencies that

differ only slightly. For example, an *amplitude modulated* radio signal is constructed by controlling the strength of a high-frequency carrier wave by means of a signal containing low frequencies only. Figure 4.5 shows, in a, a low-frequency pulse that rises and falls slowly with time; in b, a high-frequency sinusoidal carrier signal; and in c, the amplitude-modulated signal that is obtained by multiplying a by b. What happens when a wave such as c is transmitted by means of a dispersive mode?

The frequency components of the amplitude-modulated signal, c of Figure 4.5, can be obtained by separately multiplying each frequency component of the low-frequency signal by the high *carrier frequency* wave of radian frequency ω_0. Let us carry out the multiplication for a single low-frequency component of radian frequency q. We shall designate this component of the product of the modulating signal and the carrier by Q:

$$Q = (\cos qt)(\cos \omega_0 t). \tag{4.1}$$

By a trigonometric identity, this is

$$Q = \tfrac{1}{2}[\cos(\omega_0 + q)t + \cos(\omega_0 - q)t]. \tag{4.2}$$

When the two frequency components of radian frequency $\omega_0 + q$ and $\omega_0 - q$ are transmitted by means of a dispersive mode, each travels at a slightly different velocity. Let k_0 be the value of k corresponding to the frequency of the ω_0 component. Let s be the slope of the ω vs k curve at $\omega = \omega_0$. Then, approximately, k for the $\omega + q$ component has a value k_+ given by

$$k_+ = k_0 + q/s. \tag{4.3}$$

Similarly, k for the $\omega_0 - q$ component has a value k_- given by

$$k_- = k_0 - q/s. \tag{4.4}$$

After the two frequency components have traveled a distance z, the product, which was given by (4.2) at $z = 0$, is Q_z, given by

$$\begin{aligned}
Q_z &= \tfrac{1}{2}\{\cos[(\omega_0 + q)t - k_0 z - qz/s] \\
&\quad +\cos[(\omega_0 - q)t - k_0 z + qz/s]\} \\
&= \tfrac{1}{2}\{\cos[(\omega_0 t - k_0 z) + q(t - z/s)] \\
&\quad +\cos[(\omega_0 t - k_0 z) - q(t - z/s_j)]\}.
\end{aligned}$$

Again, by a trigonometric identity

$$Q_z = [\cos(\omega_0 t - k_0 z)][\cos q(t - z/s)]. \tag{4.5}$$

When we examine (4.5), we see that the carrier wave of high frequency ω_0 travels at a velocity v given by

$$v = \omega_0 / k_0.$$

However, the modulating function $\cos qt$, where q is a low frequency, travels at a rate v_g given by

$$v_g = s. \tag{4.6}$$

Moreover, this is true for all frequency components of the modulating function for which the radian frequency q is small enough so that the slope s of the ω vs k curve does not change appreciably in the frequency range between $\omega_0 - q$ and $\omega_0 + q$.

The quantity v_g, the slope of the ω vs k plot, is called the *group velocity*. It is the velocity with which an amplitude function impressed on a carrier wave travels in the z direction.

The amplitude function may be a pulse, as shown in Figure 4.5. When this is so, a wiggly pulse, as shown in c of Figure 4.5

travels in the z direction with a velocity v_g. There is no energy to the left or right of the pulse, but there is energy within the pulse. Hence, the group velocity is the velocity with which the energy of a wave travels in the z direction. If E is the energy per unit length, the power P, that is, the rate flow of energy, is

$$P = v_g E. \tag{4.7}$$

If v_g is positive (a positive slope), the energy of the wave travels to the right. If v_g is negative (a negative slope), energy travels to the left.

Figure 4.6 shows that the group velocity might be either smaller or larger than the phase velocity. It is of interest to consider some simple examples of dispersive modes.

Electromagnetic waves can travel through a metal tube or waveguide only if the frequency is greater than a *cutoff* frequency ω_0. Figure 4.7 shows the ω vs k curves for two different modes of a waveguide. The intercepts ω_{01} and ω_{02} are the cutoff frequencies for the two modes. We notice that at the cutoff frequency, $k = 0$ and the slope is zero, so the group velocity is zero. At this point, ω/k, the phase velocity, is infinite. The curves of Figure 4.7 are hyperbolas. For very large values of ω and k, the ω vs k curves approach a common asymptote, the dashed line through the origin. The slope of this line is the velocity of light. Thus, when ω and k are large, both the phase velocity v and the group velocity v_g approach the velocity of light.

A long, straight dielectric rod can guide an electromagnetic wave. At very high frequencies, most of the electromagnetic energy travels inside the rod and the electromagnetic field outside of the rod falls off rapidly with distance from the rod. In this case, the phase velocity is very close to the velocity of an electromagnetc wave in a space filled with the dielectric. At low

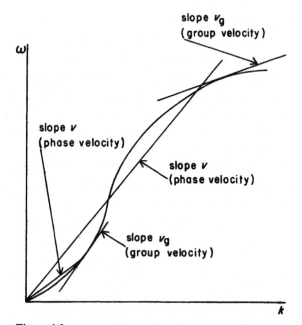

Figure 4.6

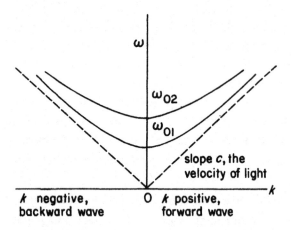

Figure 4.7

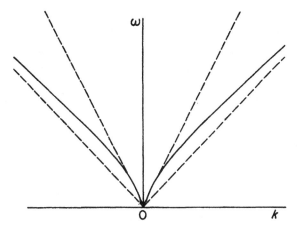

Figure 4.8

frequencies, most of the electromagnetic energy travels outside of the dielectric rod and the phase velocity is close to that of an electromagnetic wave in air or vacuum. Thus, the ω vs k curve is as shown in Figure 4.8. At very low and very high frequencies the phase and group velocities are equal; at intermediate frequencies the phase velocity is always greater than the group velocity.

We have noted that the group velocity is the velocity of energy flow. Let us imagine that a mode carries a specified amount of power, P. Equation 4.7 tells us that if the group velocity v_g is high the energy E per unit length is small; if the group velocity v_g is low the energy per unit length is great. Consider the waveguide waves illustrated in Figure 4.7. The group velocity is very small just above the cutoff frequency ω_{01}. Thus, for frequencies just above the cutoff frequency, for a given power flow the energy per unit length is high.

In Chapter 2 we expressed average power P in terms of peak strength S_0 as

$$P = S_0^2. \tag{2.14}$$

We did this by choosing the units in which the strength S is measured in such a way that the instantaneous power P_i and the instantaneous strength S are related by

$$P_i = 2S^2. \tag{2.12}$$

This is very appropriate if we consider waves of one frequency only. However, if we consider the same mode at various frequencies, we have to measure S in different units at different frequencies in order to make (2.12) hold. If we measure S in constant units, such as volts or dynes or kilometers/second, we must write

$$P_i = 2KS^2. \tag{4.8}$$

Here K is a quantity that changes with frequency. We can also express the average energy per unit, E, in terms of S_0:

$$E = AS_0^2. \tag{4.9}$$

For a given mode, if we measure S_0 in constant units, (4.8) and (4.9) hold, and K and A can change with frequency. However, in many cases S_0 is physically more nearly related to energy per unit length E than it is to power P, so, as frequency changes, K changes more than A does and (4.9) is a more enlightening relation than (4.8) is.

Let us assume that A of (4.9) remains constant, or nearly constant. Then we have

$$P = v_g E = A v_g S_0^2. \tag{4.10}$$

We can then say that for unit power ($P = 1$), S_0 is large if the group velocity is small and small if the group velocity is large.

Let us consider a medium whose properties change gradually with distance. Such a medium might be a waveguide whose diameter decreased slowly with distance. In this case, the cutoff frequency of a particular mode, ω_{01}, increases with distance. If the change is gradual enough, a wave of a given frequency travels on the medium almost as if its properties did not change with distance. How does the wave behave in so doing?

Consider a wave of unit power and of radian frequency ω which propagates in a particular mode above cutoff frequency. As the waveguide narrows, ω_{01} increases. Thus, ω/ω_{01} decreases. Thus, the group velocity decreases, and the peak strength s_0 increases as the wave travels along the tapered waveguide. In the case of the waveguide, this means that the electric field strength increases.

If the guide becomes narrow enough, ω_{01} becomes equal to ω. The particular wave or mode cannot travel beyond this point; it is reflected, so that a wave with negative velocity travels back along the guide. From (4.10) we would deduce that S_0 becomes infinite at the point where $\omega = \omega_{01}$, but this is not so. These equations are not accurate when v_g really goes to zero.

Nonetheless, when we launch a wave into a narrowing waveguide, the electric field strength does increase as the guide narrows and the cutoff frequency increases. And a similar behavior is found in other cases.

The basilar membrane of the inner ear is a transmission medium that carries waves from its broad end, at the *stapes* (near the eardrum), to its narrow, *apical* end. In a way, the behavior of the tapered basilar membrane is just the opposite of that of a waveguide. To a certain distance or place beyond the eardrum end, the basilar membrane transmits waves of all frequencies *below* a cutoff frequency ω_c. This *upper* cutoff frequency *decreases* as the width of the basilar membrane decreases. Thus, only low-frequency waves can travel all the way from the eardrum end to the narrow apical end of the basilar membrane. Waves of most frequencies are reflected part way along the basilar membrane. The basilar membrane is very lossy; in the process of reflection most of the power of the wave is absorbed, so that the wave actually reflected has negligible strength.

Figure 4.9 shows the envelope of the strength of a sinusoidal wave traveling along the basilar membrane from left to right, as given in *The Speech Chain* by Denes and Pinson.[1] The envelope is the extreme displacement at each point as the wave travels past. For a given frequency ω, the strength is greatest at the *place* of reflection. As the frequency ω is increased, this place moves toward the stapes (eardrum) end. Thus, the place of greatest wave strength depends on frequency. The frequency dependence of the place of reflection has a part in our ability to sense the frequency of sound waves.

Not long ago I saw a beautiful example of the change of wave strength with group velocity. Inside the glass facade of the New York State Theater at Lincoln Center hangs a sort of curtain made of separate strands of metal beads strung on wires or cords. During an intermission I was standing on a walkway near the very top of these strings of beads, and I idly plucked a string to see how waves would travel on it. A wave did indeed travel down the string of beads. It was barely noticeable at the top, where I plucked the string, but the displacement of the beads became greater as the wave traveled downward, and the bottom of the string swished left and right wildly. Moreover, the wave traveled up and down repeatedly, reflected at the top and the bottom ends of the string.

A stretched string of beads is an almost dispersionless medium for any waves of wavelength long compared with the

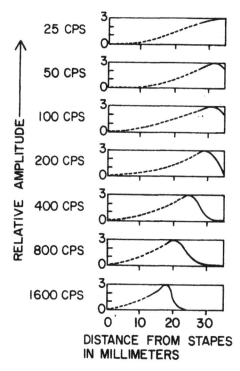

Figure 4.9 Envelope of basilar membrane displacement for different frequencies of sinusoidal excitation applied at the stapes.

distance between beads. For such waves, the group velocity and the phase velocity are equal. Waves of different frequency travel along together, so the pulse excited by plucking maintains its shape as it travels along.

The group and phase velocities of a wave of a stretched string increase as the tension increases. The tension of the strings of beads in the New York State Theater depends entirely on the weight of the beads. The tension is great at the top of the strings and zero at the bottom. Hence, the phase and group velocities are high at the top of the string and low at the bottom.

Let us ask how a disturbance on the string changes from top

Figure 4.10

to bottom. At the top the disturbance is weak (in displacement) but long. As the velocity decreases toward the bottom the transverse displacement of the beads becomes greater and the length of the disturbance becomes shorter.

I have found that the problem of waves on a hanging chain or string is an old one. Morse gives a solution in problem 9, page 149 (Chapter III) of his book, *Vibration and Sound*.[2] The solution given is a standing wave of frequency ω. Figure 4.10 shows the displacement of the string at a particular instant. The large-amplitude, short-wavelength disturbance near the bottom is easily visible; the low-amplitude, long-wavelength disturbance at the top of the string is hard to see. Both represent the same energy flow, the energy traveling with small phase and group velocity at the bottom and large phase and group velocity at the top.

References

1. Peter B. Denes and Elliot H. Pinson, *The Speech Chain*, Bell Telephone Laboratories, 1963.

2. Philip M. Morse, *Vibration and Sound*, New York: McGraw-Hill Book Company, Inc., 1948.

Problems

1. Group velocity is sometimes explained in terms of the speed at which the "beats" between two sinusoidal waves travel. Assume two equal-amplitude waves of frequency ω_0 and $\omega_0 + p$, where p is very small compared with ω. Describe the nature, and velocity of the pattern of these two waves.

2. For a mode of a waveguide

$$k = \frac{\omega}{c}\left(1 - \frac{\omega_c^2}{\omega^2}\right)^{1/2}.$$

Here c is the velocity of light and ω_c is the cutoff frequency. What are the group and phase velocities?

3. A signal $F(t)\cos(\omega t - kz)$ is sent down a waveguide, and $F(t)$ contains a range of frequencies from 0 to ω_0. What range of frequencies does the signal contain?

4. When q is large enough so that (4.3) and (4.4) are inaccurate,

we can write

$$\frac{k_+ + k_-}{2} = k_s,$$

$$\frac{k_+ - k_-}{2} = k_d.$$

Here k_+ and k_- are values of k at $\omega = \omega_0 \pm q$, k_s is the average value of k, and k_d is half the difference between the two values of k. Show that if the original signal at $z = 0$ is given by (4.1) (which is the same as (4.2)), the signal at z is given by

$$Q_z = [\cos(qt - k_d z)\cos(\omega_0 t - k_s z)].$$

Describe the effects of k_s and k_d on the signal.

5. We can expand the ω vs k curve in a power series:

$$k = k_0 + \left(\frac{\partial k}{\partial \omega}\right)_0 (\omega - \omega_0) + \frac{1}{2}\left(\frac{\partial^2 k}{\partial \omega^2}\right)_0 (\omega - \omega_0)^2$$
$$+ \frac{1}{6}\left(\frac{\partial^3 k}{\partial \omega^3}\right)_0 (\omega - \omega_0)^3 + \dots .$$

Here the subscript 0 means that the derivatives are taken at ω_0, k_0. Express k_s and k_d in terms of q by means of the power series. What terms affect the phase of the factor with a frequency q? What terms affect the phase of the factor which has a frequency ω_0?

6. In receiving and demodulating a signal the important phase is the phase with respect to a carrier of some frequency ω_0. For example, if the transmitted signal is

$$\cos \omega_0 t(1 + a \cos qt)$$

the received signal at z is

$\cos(\omega_0 t - k_0 z) + a[\cos(qt - k_d z)\cos(\omega_0 t - k_s z)]$.

If we disregard all derivatives of k higher than $\partial^2 k/\partial\omega^2$, we obtain for the received signal

$$\cos(\omega_0 t - k_0 z) + a\left\{\cos q\left[t - \left(\frac{\partial k}{\partial\omega}\right)_0 z\right]\right.$$
$$\left.\cos\left[(\omega_0 t - k_0 z) - \frac{1}{2}\left(\frac{\partial^2 k}{\partial\omega^2}\right)_0 q^2 z\right]\right\}.$$

We observe that the two cosine terms containing $\omega_0 t$ in the argument are not in phase. For what value of

$$\frac{1}{2}\left(\frac{\omega^2 k}{\partial\omega^2}\right)_0 q^2 z$$

will the terms be 90° out of phase?

7. Assume that k is as in problem 2. Show that

$$\frac{\partial^2 k}{\partial\omega^2} = -\frac{1}{c}\frac{\omega_c^2}{\omega^3}\left[1 - \left(\frac{\omega_c}{\omega}\right)^2\right]^{-3/2}.$$

8. In a proposed waveguide communication system, the waveguide is a circular pipe two inches in diameter and a mode called the circular electric or $TE_{0,1}$ mode is used. The cutoff frequency of this mode is 7.5×10^9 hertz. Assume a signal frequency of 30×10^9 hertz. What is the value of $(\partial^2 k/\partial\omega^2)_0$ (if z is measured in meters)?

9. If the waveguide in problem 8 is 30 kilometers (18.6 miles) long, for what value of q is the phase shift

$$-\frac{1}{2}\left(\frac{\partial^2 k}{\partial\omega^2}\right)_0 q^2 z$$

$-\pi/2$ radius ($-90°$). What frequency does this correspond to?

10. The expression for a phase-modulated signal is

$\cos(\omega_0 t + \phi \cos qt)$.

Show that when $\phi \ll 1$, this is approximately

$\cos \omega_0 t - \frac{1}{2}\phi[\sin(\omega_0 + q)t + \sin(\omega_0 - q)t]$.

Suppose that (because of transmission over a dispersive line) the *same* phase angle $\pi/2$ was added to the phase angles of $(\omega_0 + q)t$ and $(\omega_0 - q)t$ of each of the sine terms. What would the expression for the signal become? What sort of signal would it be?

5 Vector and Complex Representation

In Chapter 2 we considered a wave whose strength S varied as

$$S = S_0 \cos \theta. \tag{5.1}$$

Figure 5.1 shows a vector of length S_0 which makes an angle θ with respect to the horizontal axis. The projection of this vector on the axis is

$$S_0 \cos \theta.$$

Thus, when S varies as $S_0 \cos \theta$, S is given by the projection on the horizontal axis of a vector of length S_0 which makes an angle θ with respect to the horizontal axis. As θ changes with time or distance, this projection changes so as to give S.

If $t = z = 0$, then $\theta = \phi$. Thus, the direction of the vector at $z = t = 0$ is given by the *phase* ϕ of the wave.

When we compare various waves, it is convenient to think of their phases at $z = t = 0$, and to represent them by vectors making various angles ϕ with respect to the horizontal axis. (43)

By a mathematical identity,

$$S_0 \cos \phi = \operatorname{Re} S_0 e^{j\phi}. \tag{5.2}$$

Here $\operatorname{Re} S_0 e^{j\phi}$ means the real part of $S_0 e^{j\phi}$, where

$$j = \sqrt{-1}. \tag{5.3}$$

We should note that $(j)(j) = (\sqrt{-1})(\sqrt{-1}) = -1$. Another identity is

$$S_0 \sin \phi = \operatorname{Im} S_0 e^{j\phi}. \tag{5.4}$$

Here $\operatorname{Im} S_0 e^{j\phi}$ is the imaginary part of $S_0 e^{j\phi}$.

We have seen that we can describe a wave in terms of a vector

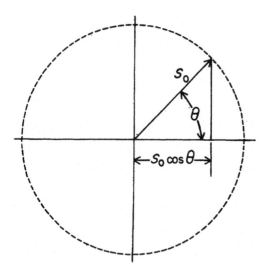

Figure 5.1

of length S_0 which makes an angle ϕ with the horizontal axis. We can regard this as a vector lying in the complex plane, as shown in Figure 5.2. The tail of the vector is at the origin. The head of the vector is at the point representing the complex number

$$e^{j\phi} = \cos \phi + j \sin \phi. \tag{5.5}$$

This point is a distance $\cos \phi$ to the right of the vertical axis and a distance $\sin \phi$ above the horizontal axis. The horizontal projection of the vector is the real part of the complex number which represents the S at $t = z = 0$; the vertical projection is the imaginary part of that complex number.

We can specify the vector of Figure 5.2 either by the peak strength or *magnitude* of the vector S_0 and its angle ϕ, or by the real and imaginary parts of the complex number that specifies the location of its head, as given by (5.2) and (5.4).

In Chapter 2 we described a sinusoidal wave of a given

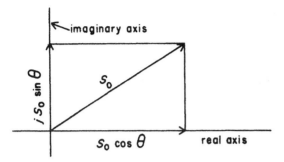

Figure 5.2

frequency and velocity by means of two numbers, the peak strength S_0 and the phase ϕ. In this section we have shown that we can descibe such sinusoidal waves by a single vector, which has both a length and direction, or by a complex number, which has a real part (represented by the horizontal component of the vector) and an imaginary part (represented by the vertical component of the vector). We choose one or another means of describing a wave as a matter of convenience or appropriateness. One representation is equivalent to the other. All involve two numerical quantities. These may be a magnitude and a phase, or two projections of a vector, or the real and imaginary components of a complex number.

The advantage of the complex representative of waves becomes apparent when certain mathematical manipulations are required.

If we desire to add two strengths S_1 and S_2 which are expressed as complex numbers, it is convenient to represent the strengths as sums of real and imaginary parts, that is, in the forms

$$S_1 = a + jb, \tag{5.6}$$

$$S_2 = c + jd. \tag{5.7}$$

The sum of S_1 and S_2 is

$$S_1 + S_2 = (a + c) + j(b + d). \tag{5.8}$$

If we desire to multiply two strengths that are represented by complex numbers, it is convenient to represent the strengths in terms of amplitudes or magnitudes, which are real numbers and phases, that is, in the forms

$$S_1 = Ae^{j\phi_1}, \tag{5.9}$$

$$S_2 = Be^{j\phi_2}. \tag{5.10}$$

Here A and B are real. The product of S_1 and S_2 is then

$$S_1 S_2 = ABe^{j(\phi_1 + \phi_2)}. \tag{5.11}$$

This is in accord with the treatment of exponents in multiplying numbers raised to various powers, as in the examples

$$(2)^2 \times (2)^3 = 4 \times 8 = 32 = 2^5,$$

$$x^n x^m = x^{n+m}.$$

We shall find later use for one interesting case of multiplication of complex numbers. Consider the complex number $e^{j(\pi/2)}$. From (5.5)

$$e^{j(\pi/2)} = \cos(\pi/2) + j \sin(\pi/2) = j. \tag{5.12}$$

We see from (5.11) that when we multiply a complex number by

$$e^{j\pi/2} = j$$

we increase its phase by $\pi/2$; that is, we rotate the vector that corresponds to the number by $\pi/2$ radius or 90° in the counterclockwise direction. If we wish, we can see that this is so by representing the original complex number as a sum of a real and an imaginary part. Let

$$W = a + jb \tag{5.13}$$

represent the strength of a wave at $z = t = 0$. The strength W can be plotted in the complex plane as shown in Figure 5.3. A line of length $|W|$ can be drawn from the origin to the point W; $|W|$ is the *magnitude* of the complex number W, and is given by

$$|W| = +\sqrt{a^2 + b^2} \ . \tag{5.14}$$

Let us consider the effect of multiplying W by j:

$$jW = ja + j(jb) = ja - b = -b + ja. \tag{5.15}$$

The complex quantity jW is also plotted in Figure 5.3.

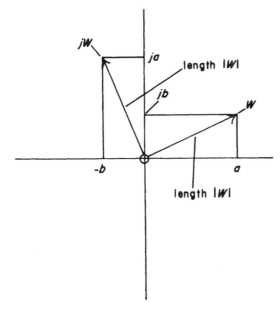

Figure 5.3

We have drawn vectors representing W and jW as lines from the origin to the points W and jW. Both have the same length $|W|$. The line from the origin to W makes the same angle with the horizontal axis as the line from the origin to jW makes with the vertical axis; hence, the lines from the origin to W and to jW are at right angles.

In complex notation, multiplication by j is a convenient way of shifting the phase or direction of the vector of a wave by $\pi/2$ radians or $90°$ in the counterclockwise direction. The magnitude of the complex number is not changed by multiplication by j.

It is easily seen that multiplication by $-j$ shifts the vector $\pi/2$ radians or $90°$ in the opposite, clockwise, direction, again without changing the magnitude.

Equation 5.14 gives an expression for the magnitude of a complex number. There is a convenient way of expressing the square of the magnitude of a complex number in terms of the number and its *complex conjugate*.

The complex conjugate of W is written as W^*. We obtain the complex conjugate of a number by replacing j with $-j$ where j occurs in the number. For instance, if

$$W = a + jb, \tag{5.16}$$

then

$$W^* = a - jb. \tag{5.17}$$

We see that

$$
\begin{aligned}
WW^* &= a^2 + jba - jab + (j)(-j)b^2 \\
&= a^2 + b^2 \tag{5.18} \\
&= |W|^2.
\end{aligned}
$$

If we represent W as

$$W = W_0 e^{j\phi},$$

then

$$W^* = W_0^* e^{-j\phi}. \tag{5.19}$$

We assume, however, that W_0 is a *real* number in which j does not appear. If W_0 is real, then

$$W_0^* = W_0$$

and

$$W^* = W_0 e^{-j\phi}.$$

We easily see that

$$WW^* = W_0^2 e^{(j\phi - j\phi)}$$
$$= W_0^2. \tag{5.20}$$

In Chapter 2 we discussed the power of a wave for which

$$S = S_0 \cos(\omega t - kz + \phi). \tag{2.1}$$

We assumed the instantaneous power P_i to be given by

$$P_i = 2S^2 = 2S_0^2 \cos^2(\omega t - kz + \phi). \tag{2.11}$$

We found that (2.11) could be written as

$$P_i = S_0^2 [1 + \cos 2(\omega t - kz + \phi)]. \tag{2.13}$$

The fluctuating term has no average value, so the average power P is

$$P = S_0^2. \tag{2.14}$$

In complex representation, the wave of (2.1) is the real part of

$$S = S_0 e^{j(\omega t - kz + \phi)}. \tag{5.21}$$

Let us now multiply S by its complex conjugate:

$$SS^* = S_0 S_0^* = S_0^2 \tag{5.22}$$

(the last because S_0 is a real number). Thus, if the strength S is expressed in complex form as in (5.21), we see from (2.14) and (5.22) that the average power P is given by

$$P = SS^*. \tag{5.23}$$

Problems

1. Two strengths are $S_1 e^{j\phi_1}$ and $S_2 e^{j\phi_2}$. Find an expression for the sum when

S_1	S_2	ϕ_1	ϕ_2
1	1	0	π
1	1	$\pi/2$	$\pi/2$
1	1	$\pi/2$	$-\pi/2$
3	4	0	$\pi/2$

2. What complex numbers S satisfy $S \cdot S \cdot S = 1$. How would you describe S?

6 Coupled Modes

Figure 6.1 illustrates a useful microwave device called a *directional coupler*. A waveguide is a metal tube through which an electromagnetic wave can travel. Two identical waveguides W_1 and W_2 have a common wall over some length L. Holes are drilled in this wall so that a wave in one waveguide interacts with or couples to a wave in the other waveguide. The result is that a fraction a of the power P_1 flowing into guide W_1 leaks into guide W_2, and a fraction $(1 - a)P_1$ remains in guide 1.

The fraction a of the power which is transferred from one waveguide to the other depends on the distance over which the waveguides are coupled and on the strength of the coupling, that is, the size of the holes. For a particular amount of coupling, $a = 1$, so that *all* of the power is transferred from one waveguide to the other.

Many other devices, including traveling wave amplifier tubes, make use of coupling of waves, that is, of modes of propagation. The consequences of such coupling can be understood by means

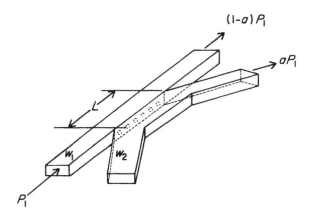

Figure 6.1

of a very simple mathematical analysis, which will be presented in this chapter.

Let us first consider the vectors that represent a wave at successive, equally spaced positions along its path of travel. Such vectors are shown in Figure 6.2. We assume that the wave is represented by a vector whose length, which we call $|S|$, is S_0. We assume that S points to the right ($\theta = 0$) at $t = 0$ and $z = z_0$. At some position z_1, a small distance from z_0 in the direction in which the wave travels, we have (again at $t = 0$)

$$\theta = \theta_1 = -k(z_1 - z_0).$$

The wave is represented by a vector of the same length, S_0, which points a little down and to the right. Again, at point z_2,

$$\theta = \theta_2 = -k(z_2 - z_0)$$
$$= -k(z_1 - z_0) - k(z_2 - z_1).$$

As we move further along the direction of travel, θ increases, and the vectors representing the wave appear as shown in Figure 6.2.

Let us now consider just what happens to S (which specifies the magnitude and phase of the wave) as we move from a point z to a point $z + dz$, where dz is a very small distance. This is illustrated in Figure 6.3. In terms of S at z, S at $z + dz$ has the same magnitude $|S|$ or S_0, but lies clockwise from S at z by a small angle. We can obtain S at $z + dz$ by adding to S at z a small vector dS which points at right angles to S at z.

In a distance dz, the angle θ which S makes with the axis must change by a quantity

$$-k\,dz.$$

In terms of the length $|dS|$ of the small vector and the length $|S| = S_0$ of the vector representing the wave, this angle is

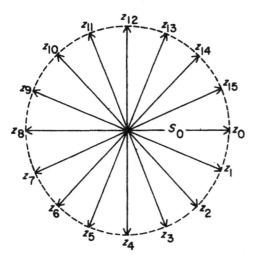

Figure 6.2

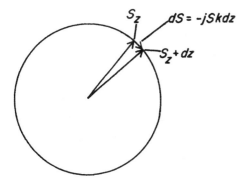

Figure 6.3

$$\frac{|dS|}{|S|}.$$

Hence

$$|dS| = Sk\,dz.$$

In complex notation, the quantity

$$Sk\,dz$$

would represent a vector of length $|S|k\,dz$ whose direction was the same as the direction of S. The little vector of length $|dS|$ which we must add to S at z in order to get S at $z + dz$ must point 90° clockwise from S. In Chapter 5, we found that we could rotate a vector 90° clockwise by multiplying it by $-j$. Hence, we can write in complex notation

$$dS = -jkS\,dz. \tag{6.1}$$

Equation 6.1 is the differential equation for the propagation of a mode. We can if we wish write it as

$$\frac{dS}{dz} = -jkS. \tag{6.2}$$

The solution of (6.2) is simply

$$S = S_0 e^{-jkz}. \tag{6.3}$$

This solution can be verified by differentiation and substitution in (6.2).

Equation 6.2 tells us that the change in S with distance is proportional to S by a constant k, just as we assumed in (6.1). When two modes are coupled, however, they interact, and the change in the strength of one is affected by the strength of the other.

Let S_1 and S_2 represent the strengths of two modes that interact through some coupling. The differential equations governing S_1 and S_2 are then

$$\frac{dS_1}{dz} = -jk_1 S_1 - jMS_2, \tag{6.4}$$

$$\frac{dS_2}{dz} = -jk_2 S_2 - jMS_1. \tag{6.5}$$

Here k_1 and k_2 are the propagation constants of the two modes in the absence of interaction, and M represents the coupling or interaction between the modes.

We may ask why S_2 and S_1 in (6.4) and (6.5) are multiplied by the *same* constant M. If we let $S_2 = DS_3$, where D is a constant, then the "coupling" constants which would multiply S_3 and S_2 would be different. We can make those constants equal by a choice of the units which we measure S_1 or S_2. The fact that k_1, k_2, and M are real quantities in (6.4) and (6.5) assures that the equations are conservative, that is, that there is no change in total power in going from z to $z + dz$.

The solutions of Equations 6.4 and 6.5 are

$$S_1 = S_{10} e^{-jkz}, \tag{6.6}$$

$$S_2 = S_{20} e^{-jkz}. \tag{6.7}$$

By differentiating (6.6) and (6.7) we obtain

$$\frac{dS_1}{dz} = -jkS_{10} e^{-jkz} = -jkS_1 \tag{6.8}$$

$$\frac{dS_2}{dz} = -jkS_{20} e^{-jkz} = -jkS_2. \tag{6.9}$$

From (6.8) and (6.9) and (6.4) and (6.5) we obtain

$$(k - k_1)S_1 = MS_2 \tag{6.10}$$

$$(k - k_2)S_2 = MS_1. \tag{6.11}$$

If we multiply the left sides together and the right sides together and divide each side by $S_1 S_2$, we obtain

$$(k - k_1)(k - k_2) = M^2, \tag{6.12}$$

$$k^2 - (k_1 + k_2)k + (k_1 k_2 - M^2) = 0. \tag{6.13}$$

The solution of the quadratic equation is

$$k = \frac{k_1 + k_2 \pm \sqrt{(k_1 + k_2)^2 + 4(M^2 - k_1 k_2)}}{2}. \tag{6.14}$$

With a little manipulation this becomes

$$k = \frac{k_1 + k_2 \pm \sqrt{(k_1 - k_2)^2 + 4M^2}}{2}. \tag{6.15}$$

When $M = 0$, that is, when there is no coupling, the $+$ and $-$ signs in (6.15) yield two values of k,

$$k = k_1,$$

$$k = k_2.$$

This is as it should be. However, if M is not zero, we also get two values of k, but these are not k_1 and k_2. Let us explore this behavior.

Let us apply our equations to the case of the two coupled waveguides in the directional coupler of Figure 6.1. In this case the phase constant of each waveguide (disregarding coupling) is the same. We shall call this phase constant in the absence of coupling k_0 so that

$$k_1 = k_2 = k_0. \tag{6.16}$$

From (6.15) and (6.16) we obtain two phase constants (corresponding to the + and − signs) which we shall call k_s and k_f:

$$k_s = k_0 + M, \tag{6.17}$$

$$k_f = k_0 - M. \tag{6.18}$$

In this example, S_1 and S_2 represent the strengths of the waves in the two guides. We can write (6.10) as

$$\frac{S_2}{S_1} = \frac{k - k_0}{M}. \tag{6.19}$$

We see that for the waves of phase constants k_s and k_f, given by (6.17) and (6.18), we have amplitude ratios S_{2s}/S_{1s} and S_{2f}/S_{1f} given by

$$\frac{S_{2s}}{S_{1s}} = \frac{k_0 + M - k_0}{M} = 1, \tag{6.20}$$

$$\frac{S_{2f}}{S_{1f}} = \frac{k_0 - M - k_0}{M} = -1. \tag{6.21}$$

In the *absence* of coupling ($M = 0$) we have two modes. One travels in guide 1 with an amplitude S_1 and a phase constant k_0. The other travels in guide 2 with an amplitude S_2 and a phase constant k_0.

When these modes are coupled (M not zero), we still have two modes. However, each is made up of a wave in both guides. One, the slower wave, has a phase constant k_s; for this mode the strengths in the two guides are equal. The other, the faster wave, has a phase constant k_f; for this mode the strengths in the two guides are equal and opposite.

Let us assume that at $z = 0$ the two waves have equal strengths, $S_0/2$, in guide 1. Then, from (6.20) and (6.21), the strengths are equal and opposite in guide 2.

In guide 1 the total strength S_{t1} is

$$S_{t1} = \frac{S_0}{2}(e^{-j(k_0+M)z} + e^{-j(k_0-M)z}). \tag{6.22}$$

In guide 2 the total strength amplitude S_{t2} is

$$S_{t2} = \frac{S_0}{2}(e^{-j(k_0+M)z} - e^{-j(k_0-M)z}). \tag{6.23}$$

We see that at $z = 0$, the total strength in guide 1 is S_0 and the total strength in guide 2 is 0. How do the strengths change with distance? We can rewrite (6.22) and (6.23):

$$S_{t1} = \frac{S_0}{2}e^{-jk_0z}(e^{jMz} + e^{-jMz}),$$
$$S_{t1} = S_0 e^{-jk_0z}\cos Mz, \tag{6.24}$$

$$S_{t2} = \frac{-jS_0}{2j}e^{-jk_0z}(e^{jMz} - e^{-jMz}),$$
$$S_{t2} = -jS_0 e^{-jk_0z}\sin Mz. \tag{6.25}$$

We see that as z increases, the total strength in guide 1 gradually decreases and the total strength in guide 2 gradually increases. When $Mz = \pi/2$, the strength in guide 1 is zero and the strength in guide 2 is jS_0, which has the same magnitude as the magnitude in guide 1 at $z = 0$. At $Mz = \pi/2$, all the power has been transferred from guide 1 to guide 2.

What if the phase constants in the two coupled guides had not been equal? In this case, coupling would indeed lead to two modes, each with some amplitude in each guide. But, for each mode the amplitudes in the two guides would not be equal in magnitude. One mode would have a phase constant near k_1, the phase constant of guide 1, and for this mode the strength in guide 1 would be larger than the strength in guide 2. The other mode would have a phase constant near k_2, the phase constant of guide 2, and for this mode the strength in guide 2 would be

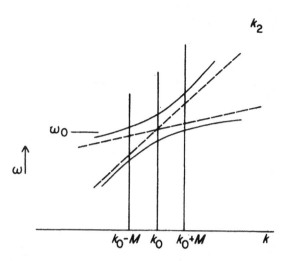

Figure 6.4

greater than the strength in guide 1. When the phase constants of the guides are different, *complete* power transfer from one guide to the other is impossible.

We have noted in Chapter 4 that the phase constants of modes can vary in different ways with frequency. Imagine that the ω vs k plot of two modes in the absence of coupling is as shown by the dashed straight lines in Figure 6.4. Now, imagine that there is some coupling between the modes. At the frequency ω_0 the phase constants k_1 and k_2 have the same value k_0. Our analysis shows that when this is so, we have two waves with phase constants $k_0 + M$ and $k_0 - M$. If we apply Equation 6.15 for other frequencies we find that the ω vs k plot for the two coupled modes is given by the solid lines of Figure 6.4. These are hyperbolas that are asymptotic to the dashed (uncoupled) ω vs k curves.

Problems

1. Assume that $k_1 = k_2$. Write expressions for the strengths S_1

and S_2 in two coupled waveguides. Show that the sum of the powers in the two guides is constant.

2. If all the power is initially in one guide, over what length L must the guides be coupled to give complete power transfer? For transferring half of the power? 1% of the power?

3. How should k_1 and k_2 be related so that if all of the power is initially in one guide, the maximum amount that is transferred to the other guide is one-half?

4. Show that power is not constant with distance in the system if M has an imaginary component.

5. Consider a device such as that shown in Figure 6.1. Assume that the coupling and length are such that exactly half the power in one mode is transferred to the other mode. Assume that we have a transmitter, a receiver, and an antenna. We want power from the transmitter to go to the antenna. We want power from the antenna to go to the receiver. We want no power to go from the transmitter to the receiver. How can we use the device to accomplish this? What will the efficiency be? Where will the rest of the power go?

7 Coupled Modes—The Other Solution

In Chapter 6 we plausibly wrote concerning coupled modes

$$\frac{dS_1}{dz} = -jk_1 S_1 - jMS_2,$$

$$\frac{dS_2}{dz} = -jk_2 S_2 - jMS_1.$$

Suppose we change a sign in one of these equations and write

$$\frac{dS_1}{dz} = -jk_1 S_1 - jMS_2, \tag{7.1}$$

$$\frac{dS_2}{dz} = -jk_2 S_2 + jMS_1. \tag{7.2}$$

The solution for k becomes

$$k = \frac{k_1 + k_2 \pm \sqrt{(k_1 - k_2)^2 - 4M^2}}{2}. \tag{7.3}$$

If

$$k_1 = k_2 = k_0, \tag{7.4}$$

then

$$k = k_0 \pm jM. \tag{7.5}$$

Equation 7.5 says that the two modes that arise through coupling vary with distance as

$$e^{-jk_0 z} e^{Mz}$$

and

$$e^{-jk_0 z} e^{-Mz}.$$

That is, one mode increases in strength as z increases, and the other decreases in strength as z increases.

This sounds implausible. We have assumed that without coupling, each of the modes is lossless. If, when we couple the modes, we obtain modes whose strengths change with distance, where does the power come from or go to?

The apparent dilemma is resolved if we imagine that in one of the original modes power is directed in one direction and in the other power is directed in the opposite direction (mathematically, the power in one original mode is positive and that in the other mode is negative). When these modes are coupled, the power that is directed to the right in one mode is gradually transferred to the other mode and is directed back to the left. Thus, the total power is zero. Figure 7.1 is a sort of schematic illustration of this. Power is directed to the right in one of the original modes and to the left in the other.

We see that the modes that result from coupling have no net power. There can be more energy per unit length where the total strength S is higher, but part of the energy is associated with power directed to the right (positive power) and part with energy directed to the left (negative power), so that there is no net power.

Such strange behavior does indeed exist in nature—in crystals, in electric filters, and in traveling-wave tubes. We shall explore this matter in subsequent chapters.

Before we go further, let us consider the nature of the ω vs k diagram for the sort of behavior we have considered. In Figure 7.2, the two dashed curves are the ω vs k curves for the uncoupled modes, for which the phase constants are k_1 and k_2. The solid curves give the phase constants when the modes are coupled.

Remember that the slope of the ω vs k curve is the group velocity v_g, the velocity with which energy of the wave travels.

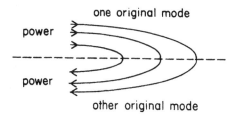

Figure 7.1

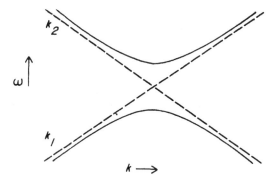

Figure 7.2

For one of the original uncoupled modes of Figure 7.2 (k_2), the slope of the ω vs k curve is negative, v_g is negative, and the energy of the wave travels to the left; for the other mode (k_1), the energy travels to the right. Hence, the conditions are just those that we have discussed, and we would expect the sort of behavior described in this chapter.

Let us examine Figure 7.2 further. For very large and for very small values of ω there are two values of k which approach k_1 and k_2, the phase constants of the modes in the absence of coupling. Thus, when k_1 and k_2 are very different, the coupling has little effect. But, for frequencies at which k_1 and k_2 are nearly equal, there is no real value of k at all. Over a range of frequencies, k is complex. The real and imaginary parts of k can be obtained from (7.3). Coupling produces two modes. The strength of one mode increases as we move to the right; the strength of the other mode decreases as we move to the right. Neither of these modes has any net power.

Reference

1. J. R. Pierce, Coupling of Modes of Propagation, *Journal of Applied Physics*, Vol. 25, pp. 179–183 (February 1954).

Problem

1. Suppose that $k_1 = -k_2 = k_0$. Describe the modes that will exist as M is gradually increased.

8 Spatial Harmonics and Coupling

Figure 8.1 shows two sine waves of wavelengths λ_1 and $\lambda_2 = \lambda_1 /2$. The wave of wavelength λ_1 travels to the right with a velocity v_1; the wave of wavelength $\lambda_1 /2$ travels to the right with a velocity $v_1 /2$.

Surely an observer can tell these two waves apart. One has a longer wavelength than the other; one travels faster than the other.

But, suppose that the observer can look at the waves only through a sequence of narrow windows W which are spaced a distance λ_1 apart. One cycle of the wave of wavelength λ_1 travels past each window in a time λ_1 /v_1. One cycle of the wave of wavelength $\lambda_1 /2$ travels past each window in a time $(\lambda_1 /2)/(v_1 /2) = \lambda_1 /v$, which is exactly the same time.

Just by looking through the windows, an observer cannot tell the long, fast wave from the short, slow wave. And, he cannot see any difference in a wave of wavelength $\lambda_1 /3$ and velocity $v_1 /3$, or in a wave of wavelength $\lambda_1 /4$ and velocity $v_1 /4$ — and on and on.

The observer who sees the waves only through the sequence of slits is blinder than this. He cannot tell in which direction any of these waves is traveling. The way in which the strength of the wave varies with time at the slits is just the same for any one of the waves going in either direction.

One consequence of these peculiar facts is that waves that "see one another," that is, that are coupled together only at a regularly spaced sequence of windows, can interact even when their wavelengths, and hence their phase constants, differ, and even if their velocities are in contrary directions. We immediately see that this is important in connection with the equations in

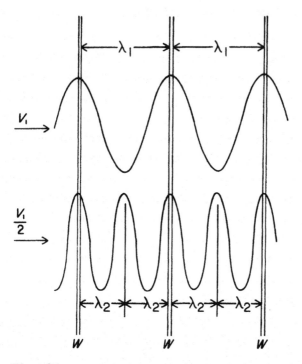

Figure 8.1

Chapter 7, which deal with the interaction of waves that carry power in opposite directions.

Let us explore this matter mathematically. Consider a sinusoidal wave of frequency ω, and phase constant k. The strength S of this wave can be written

$$S = S_0 \cos(\omega t - kz + \phi). \tag{8.1}$$

Now, suppose that we observe this wave only at narrow slits that are a distance L apart. Because the wave repeats every time $(\omega t - kz + \phi)$ changes by 2π, in going from one slit to the next we cannot tell whether the phase has changed by an amount kL between two successive slits, or by an amount $kL + 2n\pi$, where n is any integer, positive or negative. Hence, when we observe a wave through narrow slits a distance L apart, a wave with phase constant k is indistinguishable from a wave with a phase constant

$$k_n = k \pm \frac{2\pi n}{L}. \tag{8.2}$$

A wave with any such phase constant k_n exhibits the same phases at the sequence of windows that are spaced a distance L apart. Let us say that the wave *really* has a phase constant k. The other "possible" phase constants k_n are called *spatial harmonics* of the wave of phase constant k.

So far we have considered spatial harmonics to arise when we view a wave or when we couple to a wave through a sequence of narrow, equally spaced apertures. Spatial harmonics can arise in other ways. A wave exhibits spatial harmonics if it is viewed or coupled through any sort of window or aperture whose size or transparency varies periodically with distance. A wave also exhibits spatial harmonics if the medium through which it travels varies periodically with distance, so that, as the wave moves along, it alternately and regularly speeds up and slows down. In each case, if the period of variation of the coupling or the medium is L, the spatial harmonics are given by (8.2).

In the sort of cases we are considering, we deal with waves that really do have some primary phase constant k but that have (or appear to have) associated with them components with other phase constants k_n, given by (8.2). These spatial harmonic components may be considerably weaker than the main wave.

We notice in (8.2) that while k changes with frequency, $2\pi n/L$ does not change with frequency. Hence, the shape of the ω vs k curve is the same for every k_n. The energy of the wave moves along with a common group velocity, the slope of the ω vs k curve, and all the spatial harmonic components are, so to speak, carried along with the wave.

Let us now consider in more detail a very simple case. The solid lines of Figure 8.2 show the ω,k plot for a nondispersive wave. The line to the right of the origin is ω vs k for the mode that travels to the right; the line to the left of the origin is ω vs k for the mode that travels to the left. The dashed lines in Figure 8.2 represent the spatial harmonics that we obtain by adding $2\pi/L$ and $4\pi/L$, and subtracting $2\pi/L$ and $4\pi/L$ from k as given by the solid lines.

In Figure 8.3, all of the ω vs k curves of Figure 8.2, solid and dashed, are shown dashed. Figure 8.3 shows the ω vs k curves that are obtained by coupling the two modes of Figure 8.2, the mode that travels to the left and the mode that travels to the right, through their spatial harmonics. We have discussed such coupling in Chapter 7. The ω vs k curves for the modes produced by coupling are the solid curves of Figure 8.3.

The solid curves between $k = -\pi/L$ and $k = +\pi/L$ tell the whole story. The dotted curves for larger and smaller values of k merely repeat the shape of the curves in this range and add no new information.

The effect of coupling two nondispersive modes is to produce *stop bands* in various frequency ranges. Figure 8.3 tells us that in the presence of coupling there is no unattenuated transmission between pairs of frequencies $(0,\omega_1)$, (ω_2,ω_3), (ω_4,ω_5), and so on.

The approach we have used can help us to understand the presence of stop bands in complicated systems, such as a pair of coaxial helices which guide electromagnetic waves.[1]

Behavior of the general sort shown in Figure 8.3 can be of practical use. In order to make a laser, one needs a pair of mirrors that reflect light back and forth between them with very little loss. Even the reflection of a shiny silver or aluminum surface may be undesirably lossy.

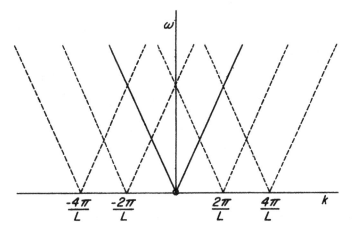

Figure 8.2

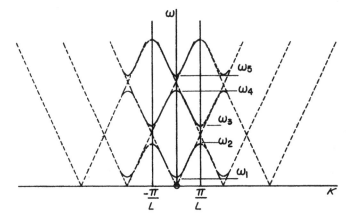

Figure 8.3

Light waves travel with different velocities in different materials. Let us choose two materials in which the velocity of light is slightly different. Let us alternate very thin, uniform layers of these materials, one on top of another. This can be done by evaporation. So we can (if we are skillful enough) obtain a material in which the velocity of light varies periodically as the light travels through it. This periodic variation of speed produces spatial harmonics.

Over certain ranges of frequency of the light, that is, in the stop bands, this layered medium will reflect light rather than transmit it. By examining Figures 8.2 and 8.3 we see that the center of such a frequency is the frequency at which $kL = \pi$. At this frequency, the pattern of layers repeats itself every half wavelength. This wavelength and the value of k in $kL = \pi$, are, of course, a sort of average value of k and a sort of average wavelength, because at a given frequency, k and λ differ slightly in the two transparent substances of which alternate layers are made.

Such a transparent, layered material makes an excellent mirror for a laser. While silver or aluminum reflects about 98% of light that falls on it, a layered surface can reflect as much as 99.9% of light that strikes it.

Further, a transparent, layered material is an *optical filter*. For, while it reflects light of some frequencies (that is, of some wavelengths, of some colors) it is transparent to light of other frequencies.

A sequence of regularly spaced discontinuities in a coaxial transmission line or in a waveguide can be used to produce a microwave filter that passes microwaves lying in some frequency ranges and rejects (reflects) microwaves lying in other frequency ranges. The analysis of this chapter gives us an idea of how such filters work, but it is not a good approach for designing microwave filters.

Physicists have found that electrons behave in some ways like particles and in some ways like waves. The waves obey an equation called *Schroedinger's equation*. According to Schroedinger's equation for an electron in a vacuum without electric or magnetic fields, the relation between k and ω is

$$\omega = \frac{h}{4\pi m} k^2. \tag{8.3}$$

Here h is Planck's constant and m is the mass of the electron:

$$h = 6.63 \times 10^{-34} \text{ joule/second};$$

$$m = 9.11 \times 10^{-31} \text{ kilogram.}$$

The form of (8.3) is shown in Figure 8.4. We see that (8.3)

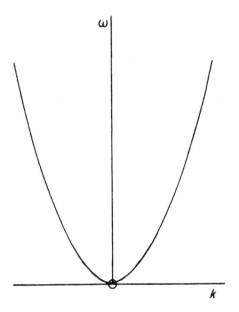

Figure 8.4

represents a dispersive wave. The phase velocity v is

$$v = \frac{\omega}{k} = \frac{h}{4\pi m}k. \tag{8.4}$$

The group velocity v_g is the slope of the ω vs k curve

$$v_g = \frac{d\omega}{dk} = \frac{h}{2\pi m}k. \tag{8.5}$$

We see that the group velocity is always twice the phase velocity:

$$v_g = 2v. \tag{8.6}$$

According to quantum mechanics, the energy is given by h times the frequency f. The radian frequency ω is equal to $2\pi f$, and ω is given by (8.3). Thus,

$$E = hf = \frac{h\omega}{2\pi} = \frac{h^2}{8\pi^2 m}k^2. \tag{8.7}$$

From (8.7) and (8.5)

$$E = \tfrac{1}{2}mv_g^2. \tag{8.8}$$

This is just the usual expression for kinetic energy, where v_g is taken as the velocity of a particle of mass m.

What about the momentum of the electron? Classically, the momentum M would be given by

$$M = mv_g = \frac{2E}{v_g}. \tag{8.9}$$

If we use (8.9) together with (8.7) and (8.5), we conclude that

$$M = \frac{h}{2\pi}k. \tag{8.10}$$

This is indeed the quantum-mechanical expression for the momentum of an electron. We see that momentum is proportional to the phase constant or wave vector k.

So far, we have considered the wave that represents an electron moving through empty field-free space. Electrons can also move freely through crystals that are regular arrays of atoms. We can get a qualitive understanding of this in terms of coupled waves. Rather than a three-dimensional array of atoms, we shall consider "flat" atoms stacked in layers a distance L apart. As electrons move through this layered medium, their waves exhibit spatial harmonics.

Figure 8.5 shows some of the spatial harmonics of Figure 8.4 as dashed curves. The solid lines show the ω vs k curves that result through coupling. Through spatial harmonics, waves representing electrons traveling to the left are coupled to waves representing electrons traveling to the right. As a result, electrons can travel freely through the "layered" crystal only if they have frequencies (that is, energies) lying in certain ranges. In Figure 8.5 the allowed energy or frequency ranges are 0 to ω_1; ω_2 to ω_3; ω_4 to some higher frequency not shown, and so on. The allowed ranges of energy in which electrons can travel freely through a perfectly regular crystal lattice are called *energy*

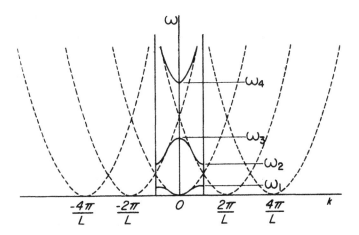

Figure 8.5

bands. Such energy bands play an important part in the functioning of such solid-state devices as transistors.

Reference

1. J. R. Pierce and P. K. Tien, Coupling of Modes in Helices, *Proceedings of the I. R. E.*, Vol. 42, pp. 1389–1396 (September 1954).

Problems

1. The strength of a wave is $S = S_0 \cos(\omega t - kz)$. We "see" this wave through a window that varies smoothly with distance so that the field strength S_1 that we see varies as $AS[1 + \cos(2\pi z/L)]$. Assuming that a second wave sees this first through the window, what values of k can the second wave have and couple strongly to the first wave?

2. In Figures 6.4 and 7.2 the solid curves that give ω vs k with coupling do not cross the dashed curves that give ω vs k without coupling. In Figures 8.3 and 8.5 the solid curves do cross the dashed curves. How do you explain this?

3. In Figure 8.3 the ω vs k curve does not touch the origin. That is, a wave will not travel on the coupled lines unless its frequency is greater than some minimum value. Is this correct? If so, how must the coupling change with frequency? What sort of coupling could this be?

9 Coupling and Tapered Media

We can apply the concept of coupling to the propagation of waves through a medium whose properties change with distance. We express the strength of the wave in such a medium in terms of the strengths of two uncoupled modes of an untapered medium, the forward mode and the backward mode. We express the effect of the changing properties of the medium as a coupling between these two modes. This coupling may vary with distance. Perhaps this approach is really a mathematical trick rather than a means for gaining physical insight. However, the approach does exhibit an interesting aspect of the equations for coupled modes.

Because we deal with one mode that has a positive group velocity and another mode that has a negative group velocity, we use the equations of Chapter 7.

In our case the wave vectors k_1 and k_2 of the two modes can be expressed

$$k_1 = -k_2 = k_0. \tag{9.1}$$

Thus, Equations 7.1 and 7.2 become

$$\frac{dS_1}{dz} = -jk_0 S_1 - jMS_2, \tag{9.2}$$

$$\frac{dS_2}{ds} = jk_0 S_2 + jMS_1. \tag{9.3}$$

Because the two waves are present in the same medium, we cannot measure S_1 and S_2 individually. Hence, it is appropriate to deal with a quantity V which is proportional to the sum of S_1 and S_2. We find that we also need a quantity W which is proportional to the difference between S_1 and S_2. Thus, we define V and W by

$$\sqrt{2}\, V = S_1 + S_2, \tag{9.4}$$

$$\sqrt{2}\, W = S_1 - S_2. \tag{9.5}$$

From (9.2)–(9.5) we obtain

$$\frac{dV}{dz} = -j(k_0 - M)W, \tag{9.6}$$

$$\frac{dW}{dz} = -j(k_0 + M)V. \tag{9.7}$$

It is appropriate to tell the reader that these equations are identical in form to the equations that govern waves on a transmission line, or in the air in a tube, or for waves on a stretched string or on a rod. They are the equations for a one-dimensional wave on any medium.

For example, consider the case of an electrical wave that travels along a transmission line whose series reactance per unit length is X and whose shunt susceptance per unit length is B. If V is the voltage across the line and I is the current in the line,

$$\frac{dV}{dz} = -jXI, \tag{9.8}$$

$$\frac{dI}{dz} = -jBV. \tag{9.9}$$

We see that (9.8) and (9.9) have exactly the same form as (9.6) and (9.7).

How do we express power in terms of V and W? We always measure S_1 and S_2 in such units that average power is $S_1 S_1^*$ or $S_2 S_2^*$. In the case of the forward and backward modes we are considering, the net power P to the right is

$$P = S_1 S_1^* - S_2 S_2^*. \tag{9.10}$$

Let us write out the quantity

$$VW^* + V^*W$$

in terms of S_1 and S_2. By means of (9.4) and (9.5) we find that

$$VW^* + V^*W = S_1 S_1^* - S_2 S_2^* \tag{9.11}$$
$$= P.$$

We should note that in the electrical case to which (9.8) and (9.9) apply,

$$P = IV^* + VI^*. \tag{9.12}$$

We see that in (9.6) and (9.7) we have equations that govern the excitation on a medium, just as (9.8) and (9.9) are the equations for the excitation on an electric transmission line. Also, V and W are two measures of the strength of the excitation, just as V and I are two measures of the strength of the excitation on an electric transmission line. From V and W we can obtain the average power by means of (9.11).

We should note that V and W need not have the same phase. Let us assume that

$$V = V_0 \cos(\omega t + \phi_v), \tag{9.13}$$

$$W = W_0 \cos(\omega t + \phi_w). \tag{9.14}$$

The instantaneous power P_i is

$$P_i = 2VW \tag{9.15}$$
$$= 2V_0 W_0 \cos(\omega t + \phi_v)\cos(\omega t + \phi_w).$$

We can make use of the following trigonometric identities:

$$\cos x \cos y = \tfrac{1}{2}[\cos(x + y) + \cos(x - y)],$$

$$\cos(x + y) = \cos x \cos y - \sin x \sin y,$$

$$\tan x = \frac{\sin x}{\cos x}.$$

By using these relationships in connection with (9.15) we can relate the instantaneous power P_i to the average power P:

$$P_i = P[1 + \cos 2(\omega t + \phi_w)$$
$$-\tan(\phi_v - \phi_w)\sin 2(\omega t + \phi_w)], \tag{9.16}$$

$$P = V_0 W_0 \cos(\phi_v - \phi_w). \tag{9.17}$$

In (9.17) the term

$$P \cos 2(\omega t + \phi_w) \tag{9.18}$$

represents a fluctuating component of the same peak amplitude as P which is *necessarily* present, as we have seen from Equation 2.13 of Chapter 2.

We find in (9.16) an additional fluctuating component of power,

$$P \tan(\phi_v - \phi_w)\sin 2(\omega t + \phi_w). \tag{9.19}$$

This component is 90° out of phase with the component given by (9.18) and thus (unless the term given by (9.19) is zero) the total fluctuation in power must be greater than the *necessary* fluctuation given by (9.18).

The additional fluctuation given by (9.19) is zero if V and W are in phase, that is, if $\phi_v = \phi_w$. If

$$\phi_v - \phi_w = \pi/2$$

there is no average power (P must be zero). The power is then entirely fluctuating power, which changes periodically from positive (to the right) and negative (to the left).

We have seen the simplicity and advantages of the complex representation of waves. Let us use complex notation and reconsider the matters we have just covered. We write

$$V = V_0 e^{j(\omega t + \phi_v)}, \tag{9.20}$$

$$W = W_0 e^{j(\omega t + \phi_w)}. \tag{9.21}$$

We have already seen in (9.11) how the power can be expressed in terms of V and W and their conjugates. Let us now consider the quantity

$$
\begin{aligned}
V^*W &= V_0 W_0 e^{-j(\omega t + \phi_v)} e^{j(\omega t + \phi_w)} \\
&= V_0 W_0 e^{-j(\phi_v - \phi_w)} \\
&= V_0 W_0 [\cos(\phi_v - \phi_w) - j \sin(\phi_v - \phi_w)] \\
&= V_0 W_0 \cos(\phi_v - \phi_w)[1 - j \tan(\phi_v - \phi_w)] \\
&= P[1 - j \tan(\phi_v - \phi_w)]. \tag{9.22}
\end{aligned}
$$

Compare (9.22) with (9.16). We see that the magnitude of the imaginary part of (9.22) is just the magnitude of the extra fluctuating component which varies with time as $\sin 2(\omega t + \phi_w)$. This is the additional fluctuating component which is not *necessarily* present in a wave whose strength varies sinusoidally with time.

The quantity

$$-jP \tan(\phi_v - \phi_w)$$

is commonly called the *reactive power*. It is a surging back and forth of energy which occurs when V and W are not in phase (or 180° out of phase).

We are now in a position to find, to investigate, and to understand the solutions of (9.6) and (9.7), for cases with and without coupling. In doing so we shall obtain a differential equation that involves V only.

First, we differentiate (9.6) with respect to z, and obtain

$$\frac{d^2 V}{dz^2} = -j(k_0 - M)\frac{dW}{dz} - j\left(\frac{d}{dz}(k_0 - M)\right)W. \tag{9.23}$$

By using (9.7) and (9.23) we obtain

$$\frac{d^2 V}{dz^2} - \frac{\dfrac{d(k_0 - M)}{dz}}{k_0 - M}\frac{dV}{dz} = -j(k_0 - M)\frac{dW}{dz}, \tag{9.24}$$

$$\frac{d^2 V}{dz^2} - \frac{d[\ln(k_0 - M)]}{dz}\frac{dV}{dz} = -j(k_0 - M)\frac{dW}{dz}$$

From (9.24) and (9.7) we obtain

$$\frac{d^2 V}{dz^2} - \frac{d[\ln(k_0 - M)]}{dz}\frac{dV}{dz} + (k_0^2 - M^2)V = 0. \tag{9.25}$$

Let us first consider the case in which the coupling M is zero and k_0 is a constant. We have

$$\frac{d^2 V}{dz^2} = -k_0^2 V. \tag{9.26}$$

The solutions of this equation are

$$V = V_0 e^{\pm jk_0 z}. \tag{9.27}$$

From (9.27) and (9.6) we obtain

$$W = \frac{j\dfrac{dV}{dz}}{k_0} = \mp V. \tag{9.28}$$

Thus, W is either in phase with V (a wave that travels in the $+z$ direction) or 180° out of phase with V (a wave that travels in the $-z$ direction). There is no reactive power.

Let us now consider a case in which k_n is a constant and

$$k_0^2 - M^2 = k_n^2, \tag{9.29}$$

$$\frac{k_0 - M}{k_n} = e^{-2Bz}, \tag{9.30}$$

$$-\ln k_n + \ln(k_0 - M) = -2Bz. \tag{9.31}$$

In this case, (9.25) becomes

$$\frac{d^2V}{dz^2} - 2B\frac{dV}{dz} + k_n^2 V = 0. \tag{9.32}$$

It is easily verified that the solution of this equation is

$$V = V_0 e^{-Bz} e^{\pm jk_n(1-B^2/k_n^2)^{1/2}z}. \tag{9.33}$$

From (9.6) and (9.30) we see that

$$W = V_0[\mp(1 - B^2/k_n^2)^{1/2} - jB/k_n]e^{Bz} e^{\pm k_n(1-B^2/k_n^2)^{1/2}z}. \tag{9.34}$$

We see that

$$V^*W = V_0^2[\mp(1 - B^2/k_n^2)^{1/2} - jB/k_n]. \tag{9.35}$$

By examining (9.35) we see that there is a constant power and a constant reactive power. The latter goes to zero as B goes to zero, that is, when $k_n - M$ changes very slowly with distance.

Further, if $B^2 > k_n^2$, V and W change exponentially with distance. No wave travels, and the power is entirely reactive.

Equations 9.29–9.35 apply to an exponentially tapered transmission line or an exponentially tapered horn (as in some loudspeakers). For waves of short wavelength (high frequency, large k_n) propagation is much the same as in an untapered medium, except that V decreases with distance and W increases with distance. However, for very long wavelengths (low frequencies, small k's) waves cannot travel through the medium. An exponentially tapered medium (an exponential horn is an example) is a high-pass medium.

Let us now consider a medium for which

$$k_0^2 - M^2 = k_n^2, \tag{9.36}$$

$$k_0 - M = \frac{A}{z^2} k_n, \tag{9.37}$$

$$\frac{d[\ln(k_0 - M)]}{dz} = -\frac{2}{z}. \tag{9.38}$$

Equation (9.25) becomes

$$\frac{d^2 V}{dz^2} + \frac{2}{z}\frac{dV}{dz} + k_n^2 V = 0. \tag{9.39}$$

It is easy to see that the following expression is a solution of (9.39):

$$V = \frac{V_0}{z} e^{\pm jk_n z}. \tag{9.40}$$

The corresponding value of W is

$$W = (V_0/A)z(\mp 1 - j/k_n z)e^{\pm jk_n z}. \tag{9.41}$$

Thus, we have

$$V^* W = (V_0^2/A)(\mp 1 - j/k_n z). \tag{9.42}$$

Equations 9.36–9.42 apply to waves in a linearly tapered medium such as a linear (conical) horn. Far from the small end, propagation is much the same as in an untapered medium, except that V decreases with distance and W increases with distance. However, there is a reactive component of power which grows without bound as we approach the small end of the horn at $z = 0$.

It is interesting to see how our results concerning linearly and exponentially tapered horns help us in understanding a particular practical problem—that of loudspeakers.

Figure 9.1 shows linearly tapered or conical horns of various internal angles θ. When θ is 180° or π radians the conical horn becomes simply a plane and the sound wave travels out radially

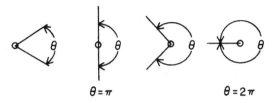

$\theta = \pi$ $\qquad\qquad \theta = 2\pi$

Figure 9.1

Figure 9.2

from the origin in all directions to the right of the plane. When θ becomes $360°$ or 2π radians the "horn" disappears and we have a spherical wave that travels from the origin equally in all directions.

The case of $\theta = \pi$ corresponds roughly to the low-frequency behavior of a cone-type speaker with an infinite plane baffle, which is illustrated at the left of Figure 9.2. The case of $\theta = 2\pi$ corresponds roughly to the case of a cone speaker in a small closed box.

In any case, we can take the radius a of the speaker aperture as corresponding to z, the distance from the origin, in (9.24). This tells us that when

$$k_n a = \frac{2\pi a}{\lambda} = 1, \tag{9.43}$$

the reactive power is equal to the true power. For wavelengths substantially longer than that specified by (9.43), the reactive power becomes very large. For such long wavelengths the speaker cone moves violently and fruitlessly in trying to radiate sound.

We noted in Chapter 1 that the speed of sound is about 1129 feet per second. Thus, if λ is measured in feet,

$$\lambda = \frac{1129}{f}. \tag{9.44}$$

When (9.42) holds, (9.43) tells us that

$$f = \frac{1129}{2\pi a} = \frac{180}{a}. \tag{9.45}$$

For a 12 inch speaker, $a = \frac{1}{2}$ foot and $f = 360$ hertz.

Actually, substantial efficiency is attained at frequencies lower than that given by (9.43) by using the springiness of the air in the closed box behind the speaker to counteract or cancel the reactive effect of the second term of (9.42) at some low frequencies. At frequencies below the resonance so attained, the power radiated falls off rapidly.

Some speakers make use of a small transducer and an exponentially tapered horn, as shown in Figure 9.3. In this case, we also encounter a reactive power at the large opening or mouth of the horn. Again, if the radius of the mouth of the horn is a, the reactive power equals the real power when (9.43) holds.

There is another limitation on the low-frequency performance of such a horn-type speaker. We see from (9.34) that there is reactive power associated with the wave within the horn. Indeed, no wave can propagate unless

$$B < k_n = \frac{2\pi}{\lambda} = \frac{2\pi f}{v} = \frac{f}{180}. \tag{9.46}$$

Thus, B is a measure of the amount of exponential taper. Consider V and W as expressed by (9.33) and (9.34). When the frequency is high and the wavelength short, the sound propa-

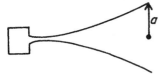

Figure 9.3

gates much as in a straight tube. The power is constant and the power density decreases as the cross sectional area of the tube horn increases. If V represents the sound pressure per unit area, we must have

V^2 (area) = constant.

This means that

$V_0^2 e^{-2Bz}$ (area) = constant

or

area = (constant) e^{2Bz}.   (9.47)

Thus,

$L = 1/2B$

is the distance in which the area increases by a factor e. In terms of this distance, condition (9.46) is

$$f > \frac{90}{L}.$$ (9.48)

Unless we observe this condition, the sound wave will not travel through the horn. In practice, we must choose L, the distance in which the area increases by a factor e, large enough so that $90/L$ is substantially less than the lowest frequency at which we wish to obtain satisfactory operation.

Problems

1. Why do you suppose the author used this strange approach rather than merely deriving the equation for a tapered transmission line?

2. Can you find any other sort of tapering for which you can solve (9.25)?

3. In the solution of (9.32) is the power the stored energy per unit length times the group velocity? Why or why not?

4. We can consider the cone of a speaker as a source of radiation of radius z which radiates into a 180° or 360° "horn." For a given z, the relation between the velocity W of the cone and the pressure V on the cone can be written (see (9.40) and (9.41))

$$W = B\left(1 - \frac{j}{p}\right)V,$$

$p = k_1 z$, which is proportional to frequency. Here B is a constant (we note that $k_1 = \omega/v$, where v is phase velocity). If there is a closed box behind the cone, the springiness of the air in the box produces another pressure on the cone, V_b, which can be written in terms of W,

$$W = BCjpV_b.$$

Here C is another constant which depends on the size of the box. The total pressure on the cone is given by $V_t = V + V_b$, and V_t/W is the acoustic impedance of the cone. Find V_t/W. In an electric analog of the acoustic system, we can let W represent current I; V be the voltage V; and R, L, and C be a resistance, an inductance, and a capacitance. Draw the equivalent circuit.

5. Is there some reason for using an exponential rather than a linear taper in a loudspeaker horn? If so, what is it?

10 Waves and Forces

We have noted that waves can exert forces on material bodies. How can we relate such forces to the energy or the power of a wave?

It is natural to start our search for such a relation by considering the case not of a moving wave but of a moving bar or a moving liquid flowing in a pipe. Let us consider a bar or a moving column of fluid of mass m per unit length which travels parallel to its axis with a velocity v_g.

The kinetic energy E per unit length is

$$E = \tfrac{1}{2}mv_g^2. \tag{10.1}$$

The power P is v_g times this quantity:

$$P = \tfrac{1}{2}mv_g^3. \tag{10.2}$$

The momentum per unit length p is

$$p = mv_g. \tag{10.3}$$

If the material that moves is a flowing liquid and if this liquid is brought to a stop as it emerges from the end of the tube by letting the emerging liquid flow into a sponge or basin, the force exerted on the sponge or basin is the momentum per unit length, p, times the velocity v_g. The length of liquid which strikes the sponge or basin in unit time is v_g. Force times time, which is now unit time, must be equal to the total momentum given up by the liquid in that time, that is, to pv_g.

The quantity pv_g is *momentum flow*. We shall call this quantity b:

$$b = pv_g = mv_g^2. \tag{10.4}$$

We see from (10.1)–(10.4) that

$$p = 2E/v_g; \tag{10.5}$$

$$b = 2P/v_g. \tag{10.6}$$

Thus, for a moving solid rod or a moving liquid, momentum per unit length p is a factor $2/v_g$ times energy E, and momentum flow b is a factor $2/v_g$ times power P. We can express forces that change E or P in terms of p or b.

We now seek useful relations between waves and forces. These relations involve quantities p and b which have the dimensions of momentum and momentum flow and which are related to forces in just the ways that momentum and momentum flow are related to forces. We shall find, however, that in some cases b and p are actually physical momentum and momentum flow, while in other cases they are not. Nonetheless, we shall call p momentum and b momentum flow, and this should cause no confusion to the forewarned reader.

Let E be the energy per unit length and P be the power of a sinusoidal wave. As we have seen,

$$P = Ev_g. \tag{10.7}$$

That is, the power is the energy per unit length times the group velocity. This follows from the fact that a pulse composed of a narrow range of frequencies travels with the group velocity v_g.

Similarly, if p is the momentum per unit length, the momentum flow b must be

$$b = pv_g. \tag{10.8}$$

Now imagine that we set up a wave by moving some device D along a dispersive medium at a speed u, from right to left, as shown in Figure 10.1. We assume that the medium is dispersive so that the group and phase velocities are different for different frequencies.

As we move the device D along, we push it to the left with a force F. We thus supply a power Fu to the wave. In addition, the device carries with it a power source that supplies a power P_1 to the wave. This power P_1 might be electric or acoustic power.

We assume that in the medium to the left of D there is no wave, and that to the right of D there is a single-frequency, sinusoidal wave traveling to the right. As the device D travels to the left, it produces this wave.

Let us now apply the conservation of energy. The stored energy between D and some fixed reference point a distance z to the right of D is

$$Ez = (P/v_g)z. \tag{10.9}$$

The rate of change of this energy is

$$(P/v_g)(dz/dt) = Pu/v_g. \tag{10.10}$$

This rate of change of energy must be equal to the power

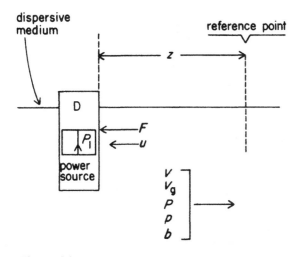

Figure 10.1

supplied through D, which is $Fu + P_1$, less the power flowing to the right past the reference point. That is,

$$Pu/v_g = P_1 + Fu - P,$$

$$P[(u/v_g) + 1] = Fu + P_1. \tag{10.11}$$

We now apply the conservation of momentum in a similar manner. The momentum between D and the fixed reference point a distance z away is

$pz.$

The rate of change of this momentum with time is

$p(dz/dt) = pu = bu/v_g.$

This must be equal to the rate at which the force supplies momentum, which is simply $-F$, less the momentum flowing past the reference point, b. Thus,

$$bu/v_g = -F - b,$$

$$b[(u/v_g) + 1] = -F. \tag{10.12}$$

We now turn our attention to the relation between energy and momentum for a solid object that moves linearly. Guided by (10.6) we assume that for a wave, b is some constant A times power. That is, we assume that

$$b = AP. \tag{10.13}$$

From (10.11), (10.12), and (10.13) we obtain

$$P[(u/v_g) + 1](1 + Au) = P_1. \tag{10.14}$$

Let us try to evaluate the constant A. Suppose we view the sinusoidal wave from the device D. If the wave travels to the left with a speed u, it will appear to be stationary when we view it

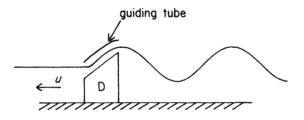

Figure 10.2

from D. For this to be so, the phase velocity v and the velocity of the device D must satisfy

$$v = -u. \tag{10.15}$$

That is, the device must move with the phase velocity of the wave.

When the wave we set up is stationary with respect to D, we can set the wave up by means of constant force, or displacement, or electric or magnetic field which moves with D. This is illustrated in Figure 10.2, in which a guiding tube sets up a wave on a dispersive medium, a medium that might be, for example, a very thin, flexible rod.

Thus, when (10.15) holds, the wave that is set up appears stationary to a person moving with D, and there is no need to supply power from a power source carried by D. The frequency as observed from D is zero and the power P_1 is zero. The only power supplied to the wave is Fu. Thus, when (10.15) holds, (10.14) becomes

$$P[(u/v_g) + 1](1 - Av) = 0. \tag{10.16}$$

This will be true if

$$A = 1/v. \tag{10.17}$$

Thus, (10.13) becomes

$$b = P/v. \tag{10.18}$$

Similarly, from (10.18), (10.7), and (10.8) we see that

$$p = E/v. \tag{10.19}$$

Relations (10.18) and (10.19) are important ones. Relation (10.18) says that the momentum flow is the power divided by the phase velocity. Relation (10.19) says that the momentum per unit length is the energy per unit length divided by the phase velocity.

We may wish to consider a signal that is a pulse made up of sine waves that have a narrow range of frequencies. In this case we can interpret (10.19) as relating a total momentum p to a total energy E.

We see that the relation between the energy and the momentum of a wave is not the same as for a solid bar that moves parallel to its axis. In the case of a moving solid bar, the momentum flow is *twice* the power divided by the *group* velocity. We must remember that the relations among power, energy, and momentum are somewhat different in the cases of waves and of material particles.

We can write (10.19) in terms of the wave vector k,

$$k = \omega/v.$$

From (10.19) we have for the momentum

$$p = (k/\omega)E. \tag{10.20}$$

Let us refer back to our discussion of Schroedinger's wave equation and to the momentum M of a particle as given in Chapter 8. There

$$M = \frac{h}{2\pi}k.$$

We also have for the energy E of a particle

$$E = \frac{h\omega}{2\pi}.$$

From the above two equations

$$M = (k/\omega)E = \frac{E}{v}. \tag{10.21}$$

This is in agreement with (10.19) and hence with (10.20).

With all this agreement, we might conclude that b and p as given by (10.18) and (10.19) are always the actual momentum flow and the momentum of a wave moving in a medium. Alas, simple cases show that this is not always true. Sometimes the actual momentum of a wave is E/v; sometimes it is not. How can this be so?

Let us review our assumptions. We have assumed that a linear wave is set up by a moving linear transducer that is acted on by a force F that produces a power Fu, and that supplies an additional power P_1 from a source that moves with the transducer. We have assumed that the force F on the moving transducer produces an actual momentum of the wave.

In order to obtain a relation between momentum and energy, we found it essential to assume that the transducer was moving, that is, the u was not zero. If we let $u = 0$, the (10.11) and (10.12) become simply

$$P = P_1, \tag{10.22}$$

$$b = -F. \tag{10.23}$$

These relations are true enough for a fixed source that sends out a wave that has a physical momentum flow b, but they cannot be used to obtain (10.18) and (10.19).

If a wave in a stationary medium generated by a stationary source has an actual physical momentum flow b, the force exerted in its generation must be equal to the actual momentum flow of the wave. Figure 10.3 shows three cases of forces exerted by waves. A source of a wave momentum flow b exerts a force $F = b$ in generating the wave. When a sink absorbs a wave of actual momentum flow b, the wave exerts on the sink a force $F = b$. If a reflector reflects a wave of actual momentum flow b as a wave of momentum flow $-b$, the incident and reflected waves exert on the reflector a force $F = 2b$.

These relations hold for a *stationary* source and a wave with

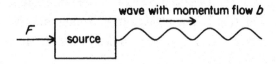

Force F of source on wave is $F=b$.

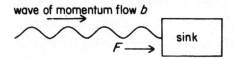

Force F of wave on sink is $F=b$.

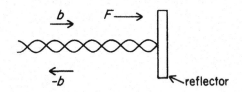

Standing wave made up of wave to the right of momentum flow b and wave to the left of momentum flow $-b$.   Force of wave on reflector is $F = 2b$.

Figure 10.3

an actual physical momentum flow b. However, when we use a moving source to set up the wave, the source can and sometimes must exert a force on the medium that supports the wave as well as on the wave itself.

We can see this most easily by considering the extreme case of a medium that consists of a linear array of uncoupled oscillators. In Figure 10.4 these are shown as circular masses M which slide frictionlessly on vertical rods. Their motion must be normal to the direction of propagation of a wave along the system, and so the masses can have no momentum in the z direction. Springs supply restoring forces, so that the masses oscillate independently with the frequency ω_0. For waves on the medium, the ω vs k diagram is as shown in Figure 10.5. The frequency is ω_0 for all values of k. The group velocity is zero; the phase velocity is ω_0/k.

We can imagine that each mass has a little projection. We can then set up a wave in the system shown in Figure 10.4 by means of a guide with a slot, shown in Figure 10.6, which engages the

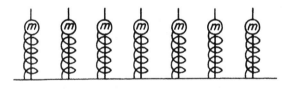

Figure 10.4

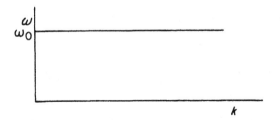

Figure 10.5

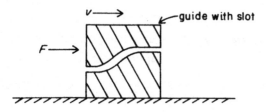

Figure 10.6

projections on the masses as it slides by at a velocity v. Clearly, some force F is necessary in order to displace the masses from their rest positions. The vertical forces exerted by the springs and vertical accelerations of the masses produce a horizontal force on the slanting slot of the guide as the guide moves along. If we call this force F, the energy E that we supply to the system when the slider moves unit distance is F. The impulse during this motion is $F/v = E/v$. This we have counted as producing a momentum E/v (per unit distance).

A wave on the medium of Figure 10.4 can have an energy E per unit length, but it can have neither power nor actual momentum in the direction of propagation, yet, relation (10.19) ascribes a momentum per unit length E/v to such a wave.

This seems very strange. In order to account for the force on a moving source, we have ascribed momentum and momentum flow to a wave that can have neither. Having ascribed such momentum and momentum flow to a wave, we should expect a force on a sink that absorbs the wave.

Even the wave in the medium of Figure 10.4 can exert a force when it is absorbed by a moving sink. The process described in connection with Figures 10.4 and 10.6 is reversible. If the guide D moves in the opposite direction, an existing wave can be destroyed. In the process, the force on the guide is correctly expressed in terms of b. But, the force is not due to physical

momentum of the wave. It is a force that is exerted on the
moving guide by the medium in the process of destroying the
wave.

A wave on the medium of Figure 10.4 cannot exert a force on
a fixed sink. It cannot even be absorbed by a fixed sink, for it
has no power flow. However, we can modify the medium of
Figure 10.4 by attaching the masses one to the next by stretched
springs; we assume that the tension in these springs remains
constant as the masses move. This medium is shown in Figure
10.7 and the ω vs k curve for waves on it is shown in Figure 10.8.

springs of constant tension

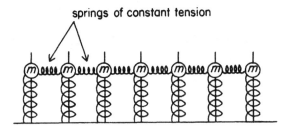

Figure 10.7

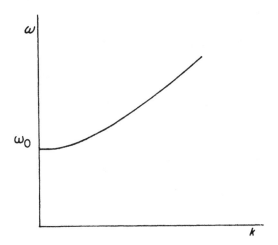

Figure 10.8

The group velocity is zero only at $k = 0$, so for other values of k the wave has power as well as energy. Thus, such a wave can be reflected by a fixed object. However, the wave can have no actual physical momentum in the z direction. Can such a wave exert a force on reflection by a fixed object?

Suppose that a wave travels from left to right on the medium of Figure 10.7. Suppose that we hold the last mass to the right fixed. The wave is reflected (a standing wave results), and the slope of the spring to the left of it changes sinusoidally with time. Because the spring is not horizontal, the average component of force toward the left due to the constant tension of the spring is less than the tension of the spring. We can interpret this reduction of the average force to the left as a pressure toward the right, a pressure produced by reflection of the wave. As a matter of fact, when the wavelength is long compared with the distance between masses, this pressure turns out to be just $2P/v$, which is consistent with the momentum flow given by Equation 10.18.

Now assume that the last mass to the right is free to move. Again the wave traveling to the right is reflected to the left, producing a standing wave. But, now the spring to the left of the mass M is essentially horizontal at all times, the force it exerts is constant, and the wave exerts no pressure on reflection.

Under some circumstances, waves without actual physical momentum can exert a force on reflection by a fixed object, but we cannot be sure what the force will be. The force is given by $2P/v$ only if P/v is actual physical momentum. The quantities, b and p as given by (10.18) and (10.19), correctly relate power, energy, and force on *moving* sources, but they attribute the force on the moving source solely to momentum of the wave generated or absorbed. The force may actually be wholly or partly a force exerted on the medium in setting up the wave.

In some cases, the force on the medium can become actual physical momentum of the medium. This would be true if the rods in Figure 10.4 were connected to a bar that could slide in the z direction without friction. Then the force exerted in setting up the wave would produce motion and momentum of the bar. Similarly, a wave in water can be accompanied by a constant velocity of the medium, and hence by momentum of a moving mass.

For electromagnetic waves in free space the momentum flow and momentum given by (10.18) and (10.19) are the actual momentum flow and momentum. This is true also for electromagnetic waves in conducting tubes of constant cross section, tubes called waveguides.

For an electromagnetic wave, the momentum density in space is given by a quantity called the Poynting vector (which gives the electromagnetic power crossing unit area) divided by c^2, where c is the velocity of light. For a guided electromagnetic wave this means that the momentum per unit length p is the power P divided by c^2:

$$p = \frac{P}{c^2}. \tag{10.24}$$

The momentum flow b is p times the group velocity:

$$b = \frac{Pv_g}{c^2}. \tag{10.25}$$

For a wave in a perfectly conducting (lossless) waveguide, the relation between k and ω is

$$k = \frac{1}{c}(\omega^2 - \omega_c^2)^{1/2}. \tag{10.26}$$

Here ω_c is a cutoff frequency; the wave cannot propagate when ω is less than ω_c.

By differentiating (10.26) we find the group velocity to be

$$v_g = \frac{\partial \omega}{\partial k} = c[1 - (\omega_c/\omega)^2]^{1/2}. \tag{10.27}$$

The phase velocity v is ω/k, or

$$v = \frac{c}{[1 - (\omega_c/\omega)^2]^{1/2}}. \tag{10.28}$$

We see from (10.27) and (10.28) that

$$\frac{v_g}{c^2} = \frac{1}{v}. \tag{10.29}$$

Hence, for the waveguide we obtain from (10.25) and (10.29)

$$b = \frac{P}{v}. \tag{10.30}$$

This is simply relation (10.18).

A waveguide of small cross section has a higher cutoff frequency than a waveguide of large cross section. Hence, at a given frequency a waveguide of small cross section has a higher phase velocity than does a waveguide of large cross section. Figure 10.9 shows a large waveguide connected to a small waveguide by a gradual taper that does not reflect the wave. We see from (10.28) and (10.30) that the momentum flow is smaller in the smaller waveguide, where the wave has a larger phase velocity than in the larger waveguide.

Clearly, the conservation of momentum requires that the wave exert a longitudinal force on the tapered section. Electric and magnetic fields can exert forces normal to perfect conductors; they cannot exert forces parallel to perfect conductors. The wave in a waveguide exerts a net outward pressure normal to the walls. In the tapered region this pressure provides the necessary component in the longitudinal direction.

Let us now consider another electromagnetic wave system, a

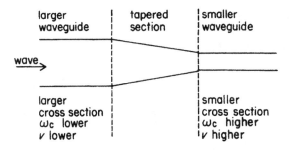

Figure 10.9

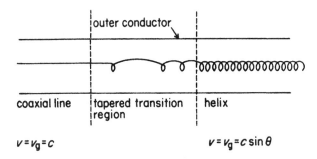

Figure 10.10

fine wire wound in the form of a helix or spring of diameter small compared to the wavelength. In such a helix, the wave travels along the wire with very nearly the speed of light. Thus, the phase and group velocities along the axis of the helix are *both* much smaller than the speed of light.

Let us imagine that the thin wire is initially straight, and is surrounded by a cylindrical outer conductor so as to form a coaxial transmission line. The wave travels along the line with a velocity c, the velocity of light. Let us further imagine that a straight perfectly conducting wire is gradually formed into such a helix, as shown in Figure 10.10.

Let $\sin \theta$ be the angle the wire of the helix makes with a plane normal to the axis. Then $\sin \theta$ is the ratio of the helix pitch to the length of one turn of the helix. Very nearly, the group and phase velocities of the wave that travels along the axis of the helix is

$$v = v_g = c \sin \theta \ll c. \tag{10.31}$$

As a wave progresses from the straight wire to the helix, the power P remains constant, so that everywhere the momentum per unit length is given by (10.24):

$$p = \frac{P}{c^2}.$$

The momentum flow b_1 in the coaxial line to the left and momentum flow b_2 in the helix to the right are, according to (10.25),

$$b_1 = \frac{Pc}{c^2} = \frac{P}{c} > \frac{P \sin \theta}{c} = \frac{Pc \sin \theta}{c^2} = b_2. \tag{10.32}$$

Thus, the wave must exert a force on the wire toward the right in the transition region. By the conservation of momentum, the total force F against the helix must be

$$F = \frac{P(1 - \sin \theta)}{c}. \tag{10.33}$$

Relation (10.18) gives a flow

$$b = \frac{P}{v} = \frac{P}{c \sin \theta}. \tag{10.34}$$

This is in disagreement with the actual momentum flow, which is given by (10.32). However, if we set up a wave by means of a moving source, relations (10.34) lead to the *total* force involved in setting the wave up. Part of the force produces momentum of the wave; part pushes against the helix.

It is interesting to contrast the waveguide of Figure 10.9 and the helix of Figure 10.10. The waveguide wall is everywhere parallel to the direction of propagation. A moving source can exert a force on the walls normal to the direction of propagation but not in the direction of propagation. Hence, any force on a moving or fixed source must produce actual momentum of the wave. In contrast, a force normal to the surface of the wire of the helix can have a component in the direction of propagation (axial direction), and hence a moving source can push against the wire in the axial direction. Thus, the force on a moving source can in part produce wave momentum and in part it can push longitudinally against the medium, that is the helix.

We have already considered cases in which a wave is set up or destroyed by a fixed or a moving source and cases in which the momentum of a wave is changed as it travels along a tapered medium. There is one case we have not considered; the coupling between one wave and another, as discussed in Chapters 6 and 7. When a wave is excited by another wave, does the wave that does the exciting count as a fixed or a moving source?

Chapters 6 and 7 say nothing about momentum, but they do give universally applicable equations for coupling between modes. Consider a case in which a wave for which P/v is actual physical momentum flow is coupled to a wave for which P/v is not actual physical momentum, a wave on the medium of Figure 10.7, for instance. Let us consider the case in which the power of the first wave is completely transferred to the second wave. Clearly, the first wave must have exerted a force P/v on something. Moreover, if the second wave transfers its power completely back to the first wave, it must give the first wave a momentum flow P/v, and so it must exert a corresponding force on the first wave. We thus conclude that when a mode on an untapered medium is coupled to a mode on another untapered medium it acts as a moving source would and the forces associated with the coupling are correctly given in terms of $p = E/v$ and $b = P/v$.

In later chapters we consider interactions with waves that travel on moving media. For such interactions we shall get the right answers if we think of b and p as given by (10.18) and (10.19) as momentum flow and momentum per unit length. For some sorts of waves, (10.18) and (10.19) are the actual momentum; for others, they are merely convenient quantities.

What we *can* say concerning b and p as given by (10.18) and (10.19) is that, when we make calculations concerning waves interacting with sources (or bodies) that move with respect to the medium in which the waves travel, we get the *right results* if we assume that the momentum flow b and the momentum per unit length p are related to the power P and the energy per unit length E by

$$b = P/v = (P/\omega)k, \qquad\qquad (10.35)$$

$$p = E/v = (E/\omega)k. \qquad\qquad (10.36)$$

In (10.36) we can consider p and E to be either momentum per unit length and energy per unit length, or, alternately as total momentum and total energy.

Problems

1. A plane electromagnetic wave enters a region in which the dielectric constant very gradually increases with distance and then maintains a constant value. Does the wave exert a longitudinal force on the medium? If so, how much?

2. Suppose that we wish to make calculations concerning the interaction of a moving source with a wave. Is it wise to use the real physical momentum of the wave if this differs from p of (10.19)? Give reasons.

3. Can a wave on the medium of Figure 10.4 be reflected by a moving source?

4. The cutoff frequency ω_c of a waveguide is proportional to the reciprocal of the width W. Consider a circular (cylindrical) waveguide that is tapered as shown in Figure 10.9. Derive the total outward force on the walls per unit length in terms of P, ω_c/ω, and W.

5. A wave on a dispersionless medium, for which the phase velocity v does not change with frequency, travels to the left and is reflected by a reflector which travels to the right with a velocity u. If the power of the wave to the left is P_1, what is the power P_2 of the reflected wave which travels to the right? The expression involves u and v. The conservation of momentum and energy may be helpful.

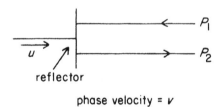

phase velocity = v

Figure 10.11

6. In problem 5, what is the ratio of the frequencies of the two waves?

11 Energy and Momentum of Waves in a Moving Medium

Let us go back to (10.14), which tells us the power P_1 that a moving source supplies to a wave in terms of P, the power of the wave, and (10.17), which gives the constant A, and (10.7) which relates the power P of the wave to the energy per unit length, E. We obtain

$$P_1 = E(u + v_g)(1 + u/v). \tag{11.1}$$

We now rewrite this equation in terms of the quantities that would be observed by a person moving with the device D shown in Figure 10.1. As a first step it is convenient to relabel certain quantities in (11.1). This we do in the following way:

replace E by E_0;

replace v_g by v_{g0}; $\hspace{4cm}$ (11.2)

replace v by v_0.

Then (11.1) is written

$$P_1 = \frac{E_0}{v_0}(u + v_{g0})(u + v_0). \tag{11.3}$$

From D we see the medium moving to the right with a velocity u. The waves travel on this medium with phase and group velocities v_0 and v_{g0}. For a fixed observer the velocities of the waves and of pulses formed by goups of waves must add, so that v and v_g must be given by

$$v = v_0 + u, \tag{11.4}$$

$$v_g = v_{g0} + u. \tag{11.5}$$

Thus, in terms of v and v_g, (11.3) becomes

$$\frac{P_1}{v_g} = \frac{E_0}{1 - u/v}. \tag{11.6}$$

The power P_1 is the power that is transferred from the fixed source to the wave on the moving medium. It is simply the power P of the wave. Hence, P_1/v_g is simply the energy E per unit length of the wave on moving medium. Thus, (11.6) can be written

$$E = \frac{E_0}{(1 - u/v)}. \tag{11.7}$$

Equation (11.7) relates the energy E per unit length of a wave on a moving medium to the energy E_0 per unit length which is observed by a person moving with the medium.

The energy E_0 is always positive. But we see that the energy of a wave on a moving medium can be either positive or negative, depending on whether the phase velocity v is greater or less than the velocity u of the medium.

How can the energy of a wave on a moving medium be negative? We shall call the energy of a wave on a moving medium negative if the medium which moves with a constant average velocity u has less energy in the presence of a wave than it does in the absence of a wave. Suppose that the wave is a longitudinal acoustic wave in a moving rod of *average* velocity u and *average* mass per unit length m. In the presence of a wave there is a small sinusoidal variation in the *actual* mass per unit length and a small sinusoidal variation in the *actual* velocity of the medium. If the actual mass per unit length is largest where and when the actual velocity, due to both the wave and average motion of the medium, is least, and if the actual mass per unit length is least where and when the actual velocity, due to both the wave and the average motion of the medium, is greatest, then the average kinetic energy in the presence of the wave is less than the average kinetic energy in the absence of the wave.

We have characterized E as given by (11.7) as the energy E per unit length of a wave on a moving medium. This must be interpreted in the same sense in which b and p have been called momentum flow and momentum.

When a fixed observer sets up a wave on a moving medium, the force he exerts is that which he would exert if b and p were actual momentum flow and momentum. In some cases they are. In other cases, the force is not exerted on the wave. It is exerted on the medium during the process of generating the wave. In this case, b and p are not the physical momentum flow and momentum of the wave. They are, however, useful quantities which have the dimensions of momentum flow and momentum.

In the same way, E as given by (11.7) is sometimes the difference between the energy of a moving medium in the presence of a wave and the energy of the same medium moving at the same average velocity in the absence of a wave. In other cases, however, the physical energy of the moving medium in the presence of the wave may be different from E of (11.7). In fact, the energy may always be greater in the presence of the wave.

This is true, for example, for a torsional wave in the medium shown in Figure 11.1. Here a number of transverse bars are strung along a thin rod or wire. When such a system moves along the axis of the rod or wire, the velocities due to the wave are normal to the average velocity of motion. Physically, the total energy in the presence of a wave must be the sum of the energy of the medium in the absence of a wave and the energy of the wave measured as if the medium were fixed. Yet, when we use a linear passive transducer in setting a wave up on this moving medium, the energy we must supply is E as given by (11.7), and this energy can be negative.

When the physical energy is not E of (11.7), part or all of the

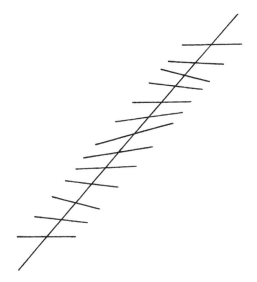

Figure 11.1

energy we expend goes into the medium rather than into the wave. Nonetheless, E of (11.7) is a very useful quantity. In practical cases a wave on a moving medium acts *as if* it had an energy E per unit length.[1-3]

Let us now turn our attention to the momentum (or what we have called the momentum) of a wave on a moving medium. We can arrive at this through (10.11), (10.12), and (10.18):

$$P[(u/v_g) + 1] = Fu + P_1;$$ (11.8)

and from (10.12) and (10.18)

$$b[(u/v_g) + 1] = -F = \frac{P}{v}[(u/v_g) + 1].$$ (11.9)

From (10.11) and (11.9)

$$-F = \frac{P_1}{(u + v)}.$$ (11.10)

By using (11.1) we obtain

$$-F = \frac{E(u + v_g)}{v}. \tag{11.11}$$

We again replace E, v, and v_g by E_0, v_0, and v_{g0} in accord with (11.2). We obtain

$$-F = \frac{E_0(u + v_{g0})}{v_0} \tag{11.12}$$

Now, the force F is directed to the left, so that the force directed to the right, the direction of travel of the wave, is $-F$. This is the only force exerted in setting up the wave. Hence, $-F$ must be equal to the momentum flow of the wave, b:

$$-F = b. \tag{11.13}$$

Accordingly, from (11.12) and (11.13)

$$b = \frac{E_0(u + v_{g0})}{v_0}. \tag{11.14}$$

The momentum per unit length p is

$$p = \frac{E_0(u + v_{g0})}{v_0 v_g}. \tag{11.15}$$

The quantity E_0/v_0 is simply p_0, the momentum per unit length when $u = 0$. The quantity $(u + v_{g0})$ is simply the group velocity v_g. Hence

$$p = p_0. \tag{11.16}$$

The momentum per unit length is unchanged by the motion of the medium. The momentum flow will of course be changed.

Let us return to (11.7), which gives the energy E as seen by a fixed observer. The energy E can be positive or negative, depending on whether v is smaller or larger than u. The phase velocity v is smaller than the velocity u of the medium only if the wave travels backward with respect to the medium, that is, if the wave travels to the left with respect to the medium. Suppose, for instance, that when moving with the medium we see forward and backward waves with phase velocities $+v_p$ and $-v_p$. Then, when the medium moves past with a velocity u, the observed phase velocities v_f and v_s are

$$v_f = u + v_p, \tag{11.17}$$

$$v_s = u - v_p. \tag{11.18}$$

The wave of velocity $u + v_p$ is a *fast* wave; the wave of phase velocity $u - v_p$ is a *slow* wave. From (11.7), the energies per unit length of the fast and slow waves, which we shall call E_f and E_s, are

$$E_f = E_0(1 + u/v_p), \tag{11.19}$$

$$E_s = E_0(1 - u/v_p). \tag{11.20}$$

We should note that when v_p is smaller than u, E_s is negative. The group velocities of both waves are positive, and so the fast wave has a positive power and the slow wave a negative power.

We have exactly the situation described above in klystrons and traveling-wave tubes. A long beam of electrons can support *space-charge waves*. If we observe these while we move along with the electrons, one space-charge wave has a phase velocity $-v_p$ (it travels to the left) and the other has a phase velocity $+v_p$ (it travels to the right). If we stand still with respect to the structure of the vacuum tube and watch the electrons move to the right with a velocity u, the phase velocities of the two space-charge waves are given by (11.19) and (11.20), and the slow wave has a negative energy per unit length and a negative power, according to (11.20). This interesting result was first given by

L. J. Chu in a paper presented at the Institute of Radio Engineers Electron Device Conference, University of New Hampshire, June, 1951. Regrettably, Chu never published his paper.

In a traveling-wave tube an electromagnetic wave travels along a helical coil of wire which surrounds an electron beam, as shown in Figure 11.2. The phase velocity of the electromagnetic wave traveling along the helix is made approximately equal to the phase velocity of the slow space-charge wave. Thus, there is coupling between the slow space-charge wave, which has negative power (energy flows to the left) and the electromagnetic wave, which has positive power. In this situation the equations of Chapter 7 apply. If the electromagnetic wave coupled to the fast, positive-power, space-charge wave, the equations of Chapter 6 would apply.

The result of the coupling is described graphically in Figure 11.3. Here u, the velocity of the electrons, is plotted vertically and the phase constant k is plotted horizontally. For the electromagnetic wave, k does not change as the electron velocity is changed, and hence the phase constant k_e of the electromagnetic wave (in the absence of coupling) appears as a vertical dashed line.

As the electron velocity u is increased, the phase constants k_f and k_s of the fast and slow waves decrease; this is indicated by the slanting dashed lines of Figure 11.3.

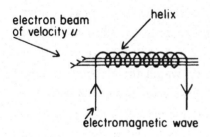

Figure 11.2

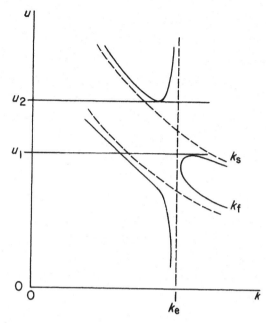

Figure 11.3

Coupling between the space-charge waves and the electromagnetic wave results in the phase constants portrayed by the
solid curves of Figure 11.3. For large and small velocities, the
effect of the coupling is small and the three phase constants are
very close to k_e, k_f, and k_s. For a range of velocities for which
the phase velocity $u - v_p$ of the slow space-charge wave is close
to the phase velocity of the electromagnetic wave, a range lying
between u_1 and u_2 in Figure 11.3, there is only one real value of
k. This real value is very nearly k_f, the phase constant of the fast
wave. The other two modes have complex phase constants,
which can be obtained from (7.3). These modes are a wave
whose strength grows exponentially with distance and a wave
whose strength decreases exponentially with distance. The power of the growing and decaying waves is zero. Each is a

combination of an electromagnetic wave component of positive power and a space-charge wave component of negative power. As the waves travel in the z direction, the positive electromagnetic power and the negative space-charge power always add up to zero.

The growing wave that we deduce from Figure 11.3 accounts for the gain of the traveling-wave tube shown in Figure 11.2. The electromagnetic wave put onto the helix excites a growing wave. As this travels along, the electron stream continually transfers power to the electromagnetic wave on the helix. At the end of the helix the increased electromagnetic power is taken off the helix to form the output of the tube. On the electron beam that emerges from the helix there is a negative-power slow space-charge wave of considerable amplitude. In other words, the power of the electron beam which emerges from the helix is less than the power of the electron beam which entered the helix. This power has been transferred to the electromagnetic wave through coupling between the electromagnetic wave and the slow space charge wave.

Growing waves can result from coupling between the space-charge waves on two electron streams which have different velocities as shown in Figure 11.4. Here it is assumed that we vary the velocity u_2 of the electrons of one electron stream. The phase constants k_{f1} and k_{s1} of the (uncoupled) waves of the other stream do not change as u_2 is changed; these are shown as vertical dashed lines. The phase constants of the other (uncoupled) stream, k_{f2} and k_{s2}, change with u_2 as shown by the dashed curves. Coupling between the modes results in the solid curves. Coupling between a fast mode and a slow mode produces and increasing wave and a decreasing wave, both with zero net power.

A single electron stream supports a fast space-charge wave

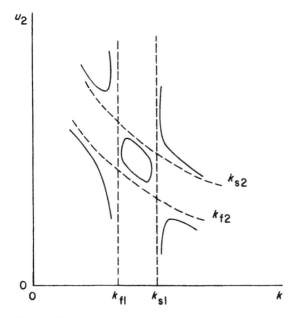

Figure 11.4

that has positive power and a slow space-charge wave that has negative power. Although these two waves have different phase constants, they can be coupled together through spatial harmonics. This can be accomplished by passing the electron beam through alternate regions of higher voltage (where they travel faster) and lower voltage (where they travel slower). Such a *velocity jump* device is shown in Figure 11.5. The effect of coupling the slow and fast space-charge waves of the single electron stream is to produce two modes, a growing wave and a decreasing wave, neither of which has any net power.

In 1949, C. C. Cutler demonstrated an effect of the general sort we have been discussing by coupling together torsional waves in a mechanical system.[4]

Torsional waves can travel along a sequence of bars fastened to a long wire or rod. This was illustrated in Figure 11.1. Cutler

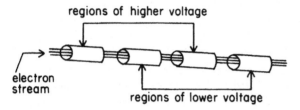

Figure 11.5

mounted two such wire-and-bar arrangements around the rims of two bicycle wheels. The bars were of steel and were magnetized slightly. The two wheels could rotate in independent directions on the same axis. The bars carried by one wheel were close enough to those carried by the other so that the waves on one wheel coupled to the waves on the other. When the wheels were rotated in opposite directions at the proper speed, a wave traveling backward on one wheel coupled to a wave traveling forward on the other wheel. Then, any small chance excitations grew rapidly in strength until there were strong waves on both wheels.

References

1. P. A. Sturrock, In What Sense Do Slow Waves Carry Negative Energy?, *Journal of Applied Physics*, Vol. 31, pp. 2052–2056 (November 1960).

2. J. R. Pierce, Momentum and Energy of Waves, *Journal of Applied Physics*, Vol. 32, pp. 2580–2584 (December 1961).

3. P. Penfield and H. A. Haus, *Electrodynamics of Moving Media*, Cambridge, Mass.: The M.I.T. Press, 1967.

4. C. C. Cutler, Mechanical Traveling-Wave Oscillator, *Bell Laboratories Record*, April 1954, pp. 134–138.

Problems

1. Think up several physical systems in which you would expect growing waves.

2. Consider a traveling-wave tube as shown in Figure 11.2. The electron beam supports two space-charge waves. The helix supports a forward wave and a backward wave. Suppose that at a given frequency we gradually increase the velocity of electrons from considerably lower than the velocity of the electromagnetic wave. Qualitatively, plot the real and imaginary parts of all k's as a function of electron velocity.

3. Three electron streams travel through the same space. One has a velocity u_1; the second has a considerably higher velocity u_2. The third has a variable velocity u. Plot qualitatively the real and imaginary components of the k's as a function of u for values ranging from less than u_1 to greater than u_2.

12 Energy and Angular Momentum in Rotating Media

In Chapter 11 we saw that waves in moving media can have, or appear to have, negative energy, and we saw how this can lead to waves that grow in strength as they travel in the direction of the moving medium. At the very end of the section we described a device built by Cutler, in which torsional waves traveled around a medium that was fastened to the rim of a rotating bicycle wheel.

Waves in such a medium, which closes on itself, are somewhat different from waves on a long, linear medium. It is just as if the wave at one end of a linear medium was forced to have exactly the same phase and strength as it had at the other end. A wave in a circular medium that is closed on itself cannot grow or decrease with distance along the medium. It can grow or decrease with time, which is just what Cutler observed when he spun the bicycle wheels in opposite directions.

When a medium is closed on itself in a circle, as shown in Figure 12.1, it is appropriate to measure position on the medium in terms of an angle θ. In terms of radius R of the medium and distance z along the medium,

$$\theta = \frac{z}{R}. \tag{12.1}$$

The strength of a wave on such a circular medium can be expressed as

$$S = S_0 \cos(\omega t - n\theta + \phi). \tag{12.2}$$

Here n plays a role analogous to the phase constant k. In fact, we can write

$$n\theta = kz = kR(z/R) = kR\theta,$$

$$n = kR = (\omega/v)R. \tag{12.3}$$

It is important to note that n must be an *integer*, positive or negative, in order for the wave to close on itself properly. This in turn means that only certain frequencies are permissible, frequencies for which kR is an integer.

A wave traveling on a medium has energy and momentum. A wave traveling on a closed, circular medium has *angular* momentum.

Imagine a particle that is a distance R from a point and that moves normal to a line between itself and the point. Let the momentum of the particle be M. This is illustrated in Figure 12.2. The angular momentum Q of the particle about the point is

$$Q = MR. \tag{12.4}$$

We note that momentum per unit length of the medium is E/v. If the medium is wound in a circle of radius R, the total angular momentum Q is

$$Q = (2\pi RE/v)R = (2\pi RE)(R/v). \tag{12.5}$$

The quantity in the first parentheses is the total energy of the wave, which we shall call W. The quantity in the second parentheses can be related to the radian frequency ω of the wave, by (12.3). We may thus write

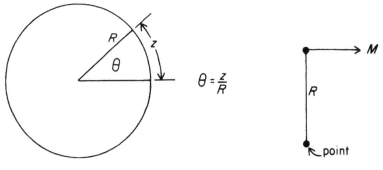

$$\theta = \frac{z}{R}$$

Figure 12.1   Figure 12.2

$$Q = \frac{Wn}{\omega}. \tag{12.6}$$

We have seen from (11.16) that the momentum per unit length is not affected by translation of the medium. Hence, if ω is the frequency observed in absence of rotation and W is the total energy observed in absence of rotation, Q is the angular momentum of the wave in the presence of any rotation.

Let us now turn to the energies of the fast and slow waves on a moving circuit, as given by (11.19) and (11.20),

$$E_f = E_0(1 + u/v), \tag{12.7}$$

$$E_s = E_0(1 - u/v). \tag{12.8}$$

As the length of the circuit is constant, the ratios of E_0/E_f and E_0/E_s are the same as the ratios of the total energies. Hence, in (12.7) and (12.8) we can, if we wish, let E_0, E_f, and E_s represent total energies rather than energies per unit length.

For a given wave, the wavelength λ does not change with the speed of the medium, but the frequency observed by a fixed observer does: it is just the observed phase velocity, $u + v_p$ or $u - v_p$, divided by the wavelength. If ω_0 is the radian frequency observed moving with the medium, and if ω_f and ω_s are the radian frequencies of the fast and slow waves, we see that

$$\omega_f = \omega_0(1 + u/v_p), \tag{12.9}$$

$$\omega_s = \omega_0(1 - u/v_p). \tag{12.10}$$

We can express u/v_p in terms of angular velocities. Let

$$\Omega = u/R, \tag{12.11}$$

$$\Omega_p = v_p/R. \tag{12.12}$$

Then

$$\omega_f = \omega_0(1 + \Omega/\Omega_p), \tag{12.13}$$

$$\omega_s = \omega_0(1 - \Omega/\Omega_p). \tag{12.14}$$

We see from (12.7) and (12.8) that

$$E_f = \frac{\omega_f}{\omega_0} E_0, \tag{12.15}$$

$$E_s = \frac{\omega_s}{\omega_0} E_0. \tag{12.16}$$

Let us now assume that an axially symmetrical stationary object supports two waves that travel about it in opposite senses. Each has a frequency ω_0, and we assume that each has a total energy E_0. What happens as we cause the object to rotate around its axis with an angular velocity Ω ?

As Ω increases, one frequency increases and the other decreases. The energies change in proportion to the frequencies. When Ω becomes greater than Ω_p, the frequency and the energy of one of the waves becomes negative. The negative frequency can be interpreted as a change in the sign of an angular velocity of the wave as observed by a stationary observer. It would perhaps be physically more appealing to say that the angular velocity of the wave changes sign when Ω becomes greater than Ω_p. The negative energy means that if a stationary device takes energy from the negative-energy wave, the strength of that wave increases.

We may note that as the earth spins on its axis, its speed at the equator is greater than the velocity of sound waves in its air or the velocity of waves on the oceans. How lucky we are that such waves cannot transfer appreciable amounts of energy to anything fixed relative to the earth. Transfer of energy through gravitational waves is too minute to matter. Transfer of energy from the upper atmosphere to the solar wind is probably too small to matter.

If sound waves or ocean waves traveling around the equator contrary to the direction of the earth's motion *could* lose energy to something other than the rotating earth they would continually increase in amplitude until nonlinearities limited their growth.

Let us now consider what happens when the two waves that travel around the object in opposite directions are coupled together. Such coupling could be provided by changes in the properties of the medium at one or more points around the circumference. Spatial harmonics would then couple the modes that travel in contrary directions.

In the sort of medium we have considered, which is wound around in a circle and closes on itself, the analog of the phase constant is the integer n of (12.2). For the spatial harmonics of such a wave, n is replaced by

$$n + m.$$

Here m is another integer, which may be positive or negative. Thus, the "phase constant" $n + m$ is an integer for all spatial harmonics. The phase constants of the spatial harmonics of two like waves that travel in opposite directions (whose phase constants are $+n$ and $-n$) are either exactly equal or considerably different.

We have seen from Chapter 7 that when we couple modes of equal and opposite phase constants, the coupling between the waves results in mixed modes in which the forward and backward, or slow and fast, waves have equal energies (for an observer who moves with the medium).

What happens from the point of view of a fixed observer? Consider Equations 12.13–12.16. First assume that

$$\Omega/\Omega_p < 1.$$

If this is so, then both E_f and E_s are positive, but E_f is greater than E_s.

Suppose we use a source of the lower radian frequency ω_s to put energy into the slow wave. In the rotating body the coupling between waves transfers some of this energy to the fast wave. To a stationary observer, the energy appears to increase in the process of transfer. We can see why this is so.

We have been discussing a situation in which some periodic unevenness or discontinuity in the medium couples a wave traveling in one direction to a wave traveling in the opposite direction and so transfers power from one wave to the other. The two waves have angular momenta in opposite directions. Hence, as power is transferred from one wave to the other, the total angular momentum of the waves, that is, the sum of the angular momenta of the two waves must change. This means that the medium must exert a force on the waves (the waves exert an equal and opposite force on the medium). Because the medium is moving, this force does work on the waves, and power is transferred from the medium to the waves.

Things work out so that if we put energy into the waves on a rotating body at the lower frequency ω_s, in order to keep the energy of the waves constant we must take a greater amount of energy out at the higher frequency ω_f. In fact, if we maintain the waves at constant strength, and if we put in a power P_s at frequency ω_s, we must take out a power P_f at frequency ω_f; P_f and P_s must be related by

$$\frac{P_f}{P_s} = \frac{\omega_f}{\omega_s}. \tag{12.17}$$

The rotating device we have described is one form of *parametric amplifier*. The moving discontinuity which causes reflection can be thought of as a parameter which changes (moves) with time, and this changing parameter is necessary for amplification to occur. The input frequency is ω_s; the output frequency is ω_f. Relation (12.17) between the input power and the output power is called the Manley-Rowe relationship. The excess of the input power over the output power is supplied by whatever keeps the rotating body rotating at an angular frequency Ω.

Suppose that

$$\Omega/\Omega_p > 1.$$

Then (12.17) holds but the ratio of P_f/P_s is negative. Power is supplied to the "source" at frequency ω_s as well as to the "load" at frequency ω_f. We have a *parametric oscillator*. We supply energy by driving the rotating body at a radian frequency Ω and we obtain power outputs at two frequencies, ω_s and ω_f.

Problem

1. Suppose we have a parametric oscillator as described toward the end of this chapter. Suppose that the medium rotates on a frictionless axle. Suppose that the modes can both lose energy to some fixed environment at a low rate but that the rate at which energy is lost increases with increasing frequency. Describe the history of the system, including the amplitudes of the modes as a function of time.

13 Parametric Amplifiers

In a traveling-wave tube, which we discussed in Chapter 11, the source of energy is a stream of moving electrons. At the output end of the tube the electrons in the stream have less energy than they had at the input end. The energy the electrons lose appears in an electromagnetic wave, which is stronger at the output end than at the input end.

Parametric amplifiers are devices in which some parameter such as capacitance, inductance, dielectric constant, or wave speed varies with time. The energy necessary to cause this variation in the presence of a signal is transferred to the signal.

In this chapter we shall consider a wave type of parametric amplifier. In such an amplifier, a wave strong enough to cause nonlinearities travels along a medium which carries other, weaker, linear waves with different but related frequencies. The changes in dielectric constant or wave speed caused by the strong wave make it possible for the strong wave to push on the weak waves so that their energy increases with distance. That increase in energy comes from the energy of the strong wave. If the strong wave is to cause the weak wave to grow continually, certain relations among the frequencies and wave vectors of the weak and strong waves must be satisfied.

In exploring the phenomena of wave-type parametric amplifiers we shall start with the idea of weak, linear waves that grow exponentially with distance. Imagine that two waves of frequencies ω_1 and ω_2 travel to the right with phase velocities v_1 and v_2. We can imagine that at a given point the average powers of the two waves are P_2 and P_1 and that these average powers increase with distance as

$$P_1 = P_{10} e^{2az}, \tag{13.1}$$

$$P_2 = P_{20}e^{2az}. \tag{13.2}$$

In a very small distance dz the changes in each power dP_1 and dP_2 are

$$dP_1 = 2aP_1\,dz, \tag{13.3}$$

$$dP_2 = 2aP_2\,dz. \tag{13.4}$$

Something must account for this change in power. We assume that something exerts an average force F per unit distance on the two waves, where

$$F = F_0e^{az}. \tag{13.5}$$

We also assume that whatever it is that pushes on the waves moves with a velocity u.

The force that acts on the waves within the distance dz is $F\,dz$. This must be equal to the increase in momentum flow within the distance dz. That is

$$\frac{dP_1}{v_1} + \frac{dP_2}{v_2} = F\,dz. \tag{13.6}$$

From (13.3)–(13.6) we have

$$2a\left(\frac{P_1}{v_1} + \frac{P_2}{v_2}\right) = F. \tag{13.7}$$

The force $F\,dz$ times the velocity u must be equal to the increase in power in the distance dz, and in a similar way this leads to

$$2a(P_1 + P_2) = Fu. \tag{13.8}$$

From (13.7) and (13.8)

$$\frac{P_2}{P_1} = -\left(\frac{1 - u/v_1}{1 - u/v_2}\right). \tag{13.9}$$

Under what circumstances can we have such an interaction between something that moves with a velocity u and the two waves? Under suitable circumstances we can have such an interaction if the waves are coupled together periodically at distances L apart, and if whatever couples the waves at these distances moves with a velocity u. When there is such coupling, any force is proportional to the momentum or momentum flow of the waves (as we have seen in Chapter 10), and hence if the power increases exponentially with distance as in (13.1) and (13.2), the force increases exponentially with distance as in (13.5).

In the continuing, additive interaction of the waves through moving, periodic coupling, two conditions must be satisfied. First, the frequencies of the two waves *as seen at the moving coupling points* must be the same. Second, the relative phases of the waves at all the coupling points must be the same, in order to make the effect of the coupling cumulative.

Let us assume that the strengths of the two waves vary with time and distance as

$$e^{az}\cos\left(\omega_1 t - \frac{\omega_1}{v_1}z\right),$$
$$e^{az}\cos\left(\omega_2 t - \frac{\omega_2}{v_2}z\right).$$

In considering frequency and phase we need concern ourselves only with the cosine terms.

The position of the coupling points can be written

$$z = ut + mL. \tag{13.10}$$

Here m is a constant and L is the distance between periodic points at which the waves are coupled. At the coupling points the cosine functions are

$$\cos\left[\omega_1 t - \frac{\omega_1}{v_1}(ut + mL)\right],$$

$$\cos\left[\omega_2 t - \frac{\omega_2}{v_2}(ut + mL)\right].$$

These may be rewritten

$$\cos\left[\omega_1(1 - \frac{u}{v_1})t - \frac{\omega_1}{v_1}mL\right], \tag{13.11}$$

$$\cos\left[\omega_2(1 - \frac{u}{v_2})t - \frac{\omega_2}{v_2}mL\right]. \tag{13.12}$$

Recall that

$$\cos \theta = \cos(-\theta).$$

Hence we may rewrite (13.12) as

$$\cos\left[-\omega_2\left(1 - \frac{u}{v_2}\right)t - \left(-\frac{\omega_2}{v_2}\right)mL\right]. \tag{13.13}$$

Expressions (13.11) and (13.12) or (13.13) give the strengths of the waves at the coupling points. The requirement that the frequencies be the same is met if the factors that multiply t in (13.11) and (13.12) or (13.13) are the same. The requirement of the same relative phase at all coupling points is met if the *difference* of the factors that multiply m is equal to $2\pi n$, where n is an integer. If this is so, the relative phase at coupling points is $2\pi mn$, which is an integral number of cycles.

Let us first use (13.11) and (13.13). The frequency condition is

$$\frac{\omega_2}{\omega_1} = -\frac{(1 - u/v_1)}{(1 - u/v_2)}. \tag{13.14}$$

This may be rewritten,

$$u\left(\frac{\omega_1}{v_1} + \frac{\omega_2}{v_2}\right) = \omega_1 + \omega_2. \tag{13.15}$$

The phase condition is

$$\frac{\omega_1}{v_1} L - \left(-\frac{\omega_2}{v_2}\right) L = 2n\pi, \tag{13.16}$$

$$\frac{\omega_1}{v_1} + \frac{\omega_2}{v_2} = \frac{2n\pi}{L}.$$

From (13.9) and (13.14) we obtain

$$\frac{P_2}{P_1} = \frac{\omega_2}{\omega_1}. \tag{13.17}$$

From (13.15) and (13.16) we obtain

$$\omega_1 + \omega_2 = n\omega_0, \tag{13.18}$$

where

$$\omega_0 = \frac{2\pi u}{L}. \tag{13.19}$$

The quantity u/L is the frequency with which coupling points pass a fixed point, and ω_0 is the corresponding radian frequency.

From (13.15) and (13.18) we can obtain a relation among phase velocities which is necessary for the operation of the amplifier:

$$\frac{\omega_1}{v_1} + \frac{\omega_2}{v_2} = \frac{n\omega_0}{u}. \tag{13.20}$$

Relations (13.16), (13.17), and (13.20) hold for a parametric amplifier in which both waves travel to the right, so that v_1 and v_2 are positive. Relation (13.17) is equivalent to relation (12.17) for the parametric amplifier described in Chapter 12.

We have noted that in actual parametric amplifers the moving coupling between waves is provided by a very strong wave of frequency ω_0, which travels through a medium that is somewhat *nonlinear*, so that the very strong wave has some effect on the weaker waves. In this case the velocity u is the phase velocity of this strong wave which *pumps* the parametric amplifier.

If the medium through which all the waves travel is dispersionless, we have

$$u = v_1 = v_2.$$

In this case (13.20) is the same as (13.18), which must hold. Thus, parametric amplification would be simple in a dispersionless medium. Unhappily, materials that make good parametric amplifiers are not dispersionless. Some special way must be found in which to satisfy (13.20).

In most actual parametric amplifiers the three waves travel in the same direction. It is possible to obtain parametric amplification when one wave travels in a direction opposite to the output wave and the pump wave of velocity u. We can obtain the equations for this case by using (13.11) and (13.12) instead of (13.11) and (13.13). In this case, the frequency condition becomes

$$\frac{\omega_2}{\omega_1} = \frac{1 - u/v_1}{1 - u/v_2}. \tag{13.21}$$

This can be rewritten

$$\omega_2 - \omega_1 = \frac{\omega_2 u}{v_2} - \frac{\omega_1 u}{v_1}. \tag{13.22}$$

The phase condition is

$$\frac{\omega_2}{v_2} - \frac{\omega_1}{v_1} = \frac{2n\pi}{L}. \tag{13.23}$$

From (13.9) and (13.21) we obtain

$$-\frac{P_2}{P_1} = \frac{\omega_2}{\omega_1}. \tag{13.24}$$

We interpret this by observing that the power P_2 of the wave traveling to the right is a positive quantity and the power P_1 of the wave traveling to the left is a negative quantity.

Equations (13.22) and (13.23) yield

$$\omega_2 = \omega_1 + n\omega_0. \tag{13.25}$$

The third relation we obtain from (13.22) and (13.25):

$$\frac{\omega_2}{v_2} = \frac{\omega_1}{v_1} + \frac{n\omega_0}{u}. \tag{13.26}$$

Equations (13.23), (13.24), and (13.25) describe a parametric amplifier in which a weak input of wave frequency ω_1 is put in from the right and travels to the left. A strong pump wave of frequency ω_0 travels to the right, and a strong output wave of frequency ω_2 is produced; this travels to the right.

In Chapter 10 we saw that in quantum mechanics, energy is proportional to frequency and momentum is proportional to wave vector or phase constant. We can if we wish think of the operation of a parametric amplifier in quantum mechanical terms. It is convenient to let $n = 1$ in this discussion. The phase constants involved are

$$k_0 = \frac{\omega_0}{u}, \tag{13.27}$$

$$k_1 = \frac{\omega_1}{v_1}, \tag{13.28}$$

$$k_2 = \frac{\omega_2}{v_2}. \tag{13.29}$$

From (13.18), (13.20), and (13.27)–(13.29), we have then for the case of three forward waves

$$\omega_1 + \omega_2 = \omega_0, \tag{13.30}$$

$$k_1 + k_2 = k_0. \tag{13.31}$$

We say that a photon of frequency ω_0 disappears and gives rise to a photon of frequency ω_1 and a photon of frequency ω_2. Equation (13.30) expresses conservation of energy and equation

(13.31) expresses conservation of momentum. A similar interpretation can be given to (13.25) and (13.26). The reader can find material related to that in this chapter in various publications.[1-3]

References

1. J. R. Pierce, Use of the Principles of Conservation of Energy and Momentum in Connection with the Operation of Wave-Type Parametric Amplifiers, *Journal of Applied Physics*, Vol. 30, pp. 1341–1346 (September 1959).

2. H. A. Haus, Power-Flow Relations in Lossless Nonlinear Media, *I.R.E. Transactions on Microwave Theory and Techniques*, Vol. MTT-6, pp. 317–324 (July 1958).

3. Amnon Yariv, *Introduction to Optical Electronics*, New York: Holt, Rinehart, and Winston, Inc., 1971, Chapter 8.

Problem

1. Think up several physical forms of parametric amplifiers.

14 Polarization

In some waves the strength of the wave has a direction as well as a magnitude. This is not true of the pressure of a sound wave in air. It is true of the vibration of a stretched string, which we first discussed in Chapter 1.

There we disregarded the fact that a stretched string can vibrate in one of two planes. In a guitar, it can vibrate parallel to the sounding board, or normal to the sounding board. Actually, there is usually some vibration in both directions.

Figure 14.1 depicts various *polarizations* of *sinusoidal waves* traveling along a stretched string; part a shows a *vertically polarized* wave, in which the string moves up and down; b shows a *horizontally polarized* wave of the same strength, in which the string moves right and left.

Of the waves shown in a and b of Figure 14.1, the horizontally polarized wave is ahead of the vertically polarized wave by a quarter wavelength; when the strength of the vertically polarized wave is at its maximum, the strength of the horizontally polarized wave is zero, and the strength of the horizontally polarized wave reaches its maximum after that wave has traveled a quarter of a wavelength.

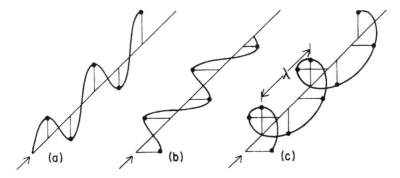

Figure 14.1

Small vibrations of a stretched string are nearly linear, so that two waves can travel on a string without affecting one another. Part c of Figure 14.1 shows the displacement of a string when the waves of a and b are both present. The strength of the wave, that is, the displacement of the string, is helical. The string is everywhere the same distance from the axis, and it winds once around the axis in the wavelength λ. In the linearly polarized waves of a and b, the strength varies with time; in the wave of c the strength is constant in magnitude and rotates with time. The wave of c of Figure 14.1 is called a *circularly polarized* wave.

If we watch the wave of c of Figure 14.1 go past a plane normal to the direction of travel, we see that the strength, measured by the point at which the string intersects the plane, will travel around the axis in a circle, moving in a *clockwise* or *right-handed* direction. If we examine the helical form of the string, we see that this corresponds to what is called a *left-handed* screw or thread. Thus, if the helix were rigid or stationary, and had a handle on it, we would have to turn the helix in a counterclockwise or left-handed direction to screw it into a cork.

If the wave of b had been a quarter wavelength *behind* that of a instead of a quarter wavelength *ahead* of the wave of b, the intersection of the wave with a plane normal to the direction of propagation would have moved around a circle in a *counterclockwise* or *left-handed* direction, and the helix itself would have been in the form of a *right-handed* screw, like that of a ordinary right-handed corkscrew.

Circularly polarized waves have either a *right-handed polarization* or a *left-handed polarization*, which is defined by convention. The TELSTAR satellite sent out circularly polarized microwaves. When it first passed over the Atlantic, the British station at Goonhilly and the French station a Pleumeur Bodou both tried to receive its signals. The French succeeded, because their definition of sense of polarization agreed with the American definition. The British station was set up to receive the wrong polarization because their definition of sense of polarization was contrary to our definition. Quite arbitrarily, we will call the wave of c of Figure 14.1 a *right-handed circularly polarized wave.*

It is of some interest to describe the waves of Figure 14.1 mathematically. If y is measured upward and x to the right, we have for the two component waves

$$y = S_0 \sin(\omega t - kz)$$
$$= S_0 \cos(\omega t - kz - \pi/2), \tag{14.1}$$

$$x = S_0 \cos(\omega t - kz). \tag{14.2}$$

For the overall wave of c, which is the sum of the waves of a and b, the radius r from the x axis is

$$r = (x^2 + y^2)^{1/2}$$
$$= [S_0^2 \cos^2(\omega t - kz) + S_0^2 \sin^2(\omega t - kz)]^{1/2} \tag{14.3}$$
$$= S_0.$$

In Chapter 2 we computed the power of a sinusoidal wave by squaring its instantaneous strength. We found the power to have an average value and a fluctuating component with twice the frequency of the wave. We should note that $x^2 + y^2$ in (14.3) is proportional to the sum of the powers of the two linearly polarized component waves, and as $x^2 + y^2$ does not vary with time, the power of a circularly polarized wave is constant with time. The average power is the instantaneous power, because the amplitude of the wave does not vary with time or distance.

A sort of analog of polarized waves is used in the transmission of electric power. In a *single phase* transmission line, there is one voltage V with respect to ground, whose time variation is

$V = V_0 \cos \omega t.$

The power transmitted necessarily fluctuates with time.

In *three phase* transmission there are three voltages with respect to time, whose time variations are

$V_1 = V_0 \cos \omega,$

$V_2 = V_0 \cos(\omega t + 2\pi/3),$

$V_3 = V_0 \cos(\omega t + 4\pi/3).$

The sum of the squares of these three voltages is constant.

Let us return to the circularly polarized wave described by (14.1)–(14.3). The angle θ that a point on the wave makes with respect to the horizontal plane (x axis) is

$$\theta = \tan^{-1} \frac{S_0 \sin(\omega t - kz)}{S_0 \cos(\omega t - kz)} \tag{14.4}$$

$$= \omega t - kz.$$

Suppose that (14.1) held but that the phase of the other wave was changed by π radians (half a wavelength), so that instead of (14.2) we had

$$x = S_0 \cos(\omega t - kz - \pi) = -S_0 \cos(\omega t - kz). \tag{14.5}$$

This results in a *left-handed circularly polarized* wave. Equation (14.3) holds, but (14.4) is replaced by

$$\theta = -(\omega t - kz). \tag{14.6}$$

In many media, waves travel in the same way no matter how they are polarized. This is not true of some materials. There are some media with magnetic properties through which an electromagnetic wave can travel with little loss. When a uniform magnetic field is present in such a material in the direction the wave travels, the speed of travel of the wave is different for left-handed and right-handed circularly polarized waves. For certain magnetic fields, the loss is different for the two circular polarizations. Thus, the medium may be transparent for one circular polarization and opaque for the other.

Under such circumstances, the absorption depends on the sense in which the angle θ, which measures the direction of the wave strength, changes with time. The angle θ changes in the same sense for a right-handed wave traveling in one direction and a left-handed wave traveling in the opposite direction. Thus, a medium can be transparent to a left-handed wave traveling in one direction and opaque to a left-handed wave traveling in the opposite direction. This makes it possible to construct one-way devices called *isolators* that transmit electromagnetic waves from the input end to the output end, but absorb any wave put into the output end.

Some media that transmit waves have different properties in different directions. Such media or materials are called *birefringent* media or materials. They exhibit different phase velocities for plane waves with polarizations at right angles to one another.

A long springy bar such as that illustrated in Figure 14.2 is birefringent. Such a bar is a dispersive medium. For a given

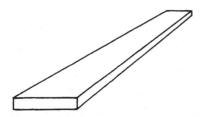

Figure 14.2

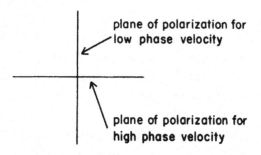

plane of polarization for
low phase velocity

plane of polarization for
high phase velocity

Figure 14.3

frequency, the phase velocity depends on the mass per unit length and on the stiffness of the bar. The bar of Figure 14.2 is clearly stiffer for bending in the horizontal direction, in which it is wider, than for bending in the vertical direction, in which it is narrower. In such a bar, for a given frequency a horizontally polarized wave travels faster than a vertically polarized wave.

In general, for a birefringent medium there are two perpendicular planes, as shown in Figure 14.3: a plane of polarization of higher phase velocity v_f and a plane of polarization of lower phase velocity v_s. If we put onto the medium a plane polarized wave that is polarized in the high velocity direction, the wave travels with a velocity v_s. But what happens to a wave polarized in any other way?

We have seen that a circularly polarized wave can be regarded as made up of two plane polarized waves. Similarly, the plane polarized wave indicated in Figure 14.4, polarized along the slanting line and having a strength S_0, can be regarded as the sum of two components, of a vertically polarized wave of strength S_{0y} and a horizontally polarized wave of strength S_{0x}. In a birefringent medium, the component that is polarized in the plane of highest phase velocity travels with a velocity v_f and the component polarized in the plane of low phase velocity travels with the velocity v_s. Thus, as the two components travel, their

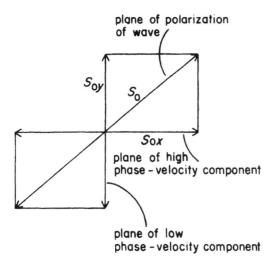

Figure 14.4

relative phase changes. This has startling consequences.

Suppose that we send in a plane-polarized wave that lies in a plane midway between right and up, a plane that makes an angle of 45° with respect to the (vertical) plane of low and (horizontal) plane of high phase velocity. In this case the high velocity and low velocity components have equal strengths. After some distance of travel, which we shall call L, the horizontal component is a quarter wavelength ahead of the vertical component. But, this is just the situation shown in Figure 14.1. At this point the plane-polarized wave has been converted into a right-handed circularly polarized wave.

We have seen that if we shift one plane-polarized component of a circularly polarized wave by half a wavelength, the direction of circular polarization is reversed. Hence, as the components travel a distance $2L$ from the point where we have a right-handed circularly polarized wave, they will constitute a left-handed circularly polarized wave. The sequence of polarizations is:

Distance	Polarization
0	linear, 45° with respect to x axis
L	circular, right-handed
$2L$	linear, 135° with respect to x axis
$3L$	circular, left-handed

and so on . . . .

A birefringent medium makes it possible to convert between linear polarization and circular polarization, and between right-handed and left-handed circular polarization. This is very useful in optical and microwave devices.

Some birefringent media are "natural" materials such as crystals of Iceland spar. Other birefringent media are fabricated. For example, a slightly elliptical waveguide has a higher phase velocity for an electromagnetic wave with an electric vector in the long direction than it does for an electromagnetic wave with its electric vector in the short direction (see Figure 14.5). Hence, a birefringent medium can be produced by slightly "squashing" a circular waveguide.

Figure 14.6 illustrates an isolator made up by using sections of a squashed circular waveguide as birefringent media, together with a material that, in the presence of a magnetic field, transmits circularly polarized waves for which θ changes with time in one sense and absorbs circularly polarized waves for which θ changes with time in the opposite sense.

Figure 14.7 illustrates a strange device that makes use of circular polarization. A squashed waveguide converts the plane polarized wave at a to a right-handed circularly polarized wave at b. The squashed center section between b and c rotates in the sense shown; it converts the right-handed circularly polarized wave at b to a left-handed circularly polarized wave at c. The

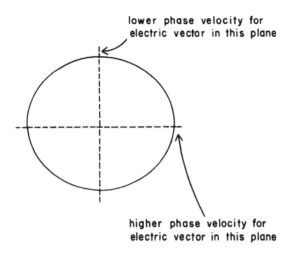

Figure 14.5

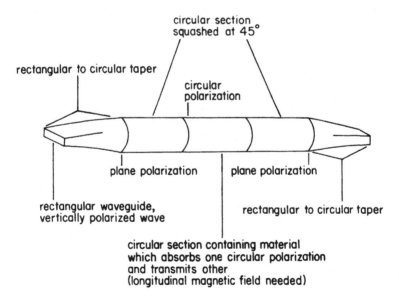

Figure 14.6

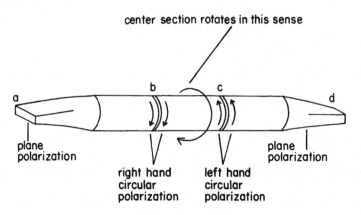

Figure 14.7

squashed waveguide between c and d converts this to a plane polarized wave at d.

Suppose we sit on the center section to the right of b and watch the circularly polarized wave come in. We are rotating in the same direction as the wave strength, so for us the strength rotates more slowly than it does for a fixed observer. In other words, for an observer moving with the center section, the frequency of the wave is lower than that observed by a fixed observer at a.

Now consider an observer just to the right of c. Let us regard the center section as fixed and this observer as rotating. He is rotating with respect to the center section in the same direction as the strength of the incoming wave. Hence, he observes a frequency lower than that observed by an observer on the center section.

We are forced to conclude that the frequency of the wave emerging at d is lower than the frequency of the wave entering at a. This is correct. If ω is the radian frequency of the wave at a and ω_0 is the angular velocity of the center section, the radian frequency of the emerging wave is $\omega - 2\omega_0$.

The device of Figure 14.7 has been used in microwave measurements at millimeter wavelengths. One part of the milli-meter-wave output went through the device to be measured; another stronger part went through the device of Figure 14.6. The two waves so obtained were fed to a nonlinear diode. The output of the diode contained a component of frequency $2\omega_0$, whose amplitude was proportional to the strength of the wave coming from the device to be measured. This component was amplified in a narrow-band amplifier.

Electromagnetic waves that travel through circular wave-guides, and transverse waves that travel through rods, and many other waves as well, can have strengths that vary as

$$S = F(r)\cos(\omega t - n\theta - kz). \tag{14.7}$$

Here $F(r)$ is some function of distance from the axis. The angle θ around the axis is measured looking along the axis in the direction in which the wave travels. We see that at a given distance from the axis the strength is greatest when

$$\theta = \frac{\omega t}{n}. \tag{14.8}$$

When n is a positive integer, (14.7) is a right-handed circularly polarized wave; when n is a negative integer, the wave is a left-handed circularly polarized wave.

In (14.7) we have a wave that has momentum in the direction of travel, as in Chapter 10, and angular momentum about the axis, as in Chapter 12. If the energy per unit length is E, we have from (10.19) that the momentum per unit length p is

$$p = \frac{E}{v} = \frac{Ek}{\omega}.$$

From (12.6) we can write q, for the angular momentum per unit length

$$q = \frac{En}{\omega}. \tag{14.9}$$

Plane polarized waves are obtained if we add two waves of the form of (14.7) if $F(r)$ is the same for each wave but n is $+m$ for one wave and $-m$ for the other (when m is an integer).

Our approach to birefringence has been that of considering the relations of conversion among circular and linear polarizations. Birefringence has other uses. We have noted in Chapter 13 that in a parametric amplifier we must satisfy the relation

$$k_1 + k_2 = k_0, \tag{13.31}$$

where the k's are the wave vectors corresponding to ω_1, ω_2, and ω_0, and

$$\omega_1 + \omega_2 = \omega_0. \tag{13.30}$$

Relation (13.30) is automatically satisfied if the medium is nondispersive, with a phase velocity v, for then (13.31) is simply

$$\frac{\omega_1}{v} + \frac{\omega_2}{v} = \frac{\omega_0}{v},$$

and this is the same as (13.30). How sad, however, if we have a dispersive medium.

Figure 14.8 shows a ω vs k plot, and three frequencies which satisfy (13.30). Here k_0 is the value of k corresponding to ω_0. We know that (13.31) would be satisfied if k_1 and k_2 lay on the dashed line through the origin. We see that if k_2 lies a certain distance to the right of this line and if (13.31) is to be satisfied, k_1 must lie an equal distance to the left, and the ω vs k curve must have a very particular S shape as shown by the solid line of Figure 14.8; this is most unlikely.

How fortunate, then, if the nonlinear medium used in a parametric amplifier is birefringent and has *two* ω vs k curves for the two plane polarizations, as shown by the solid curves of

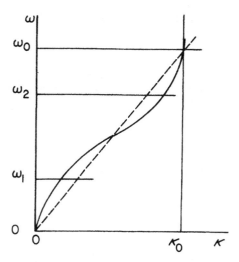

Figure 14.8

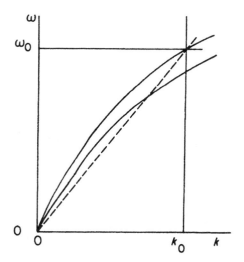

Figure 14.9

Figure 14.9. We can polarize the strong wave of frequency ω_0, which is used to couple the two weak waves, so that k_0 is given by the curve to the left. Then there is at least a chance that we can find two frequencies ω_1 and ω_2 and corresponding k's on the curve to the right, for which (13.30) and (13.31) are both satisfied.

Problems

1. List several sorts of polarized and unpolarized waves.

2. For a medium such as that shown in Figure 14.9, suggest a procedure for finding all possible combinations of operating frequencies.

15 Plane and Nearly Plane Waves

So far, we have discussed waves that are guided and confined by a medium that extends in the z direction. Such a medium may be stretched string, a metal rod, a transparent fiber that guides light, or a pair of wires or a tube which guides electromagnetic waves. We know, however, that waves can travel through open spaces. Sound waves travel through the air in a room, and light waves travel both through air and through the vacuum of space.

The term *plane wave* really can have two meanings. In a broad and strict sense, any wave whose phase is the same over a plane normal to the direction of propagation is a plane wave, even if the strength of the wave varies within that plane. Here we shall use *plane wave* rather to describe a wave that travels in some particular direction and whose strength and phase are constant over any plane normal to this direction of propagation. Such a wave might be a plane polarized or a circularly polarized electromagnetic wave. Or, it might be a sound wave in the air, which has no polarization.

Of course, such plane electromagnetic waves and plane sound waves do not actually exist in nature, but there are many waves that are so nearly plane waves that we can regard them as such. Far away from a sound source or a radio transmitter the phase is actually constant over the surface of a sphere. Thus, the waves we encounter in nature are usually spherical waves rather than plane waves. However, far from its center, a portion of the surface of a sphere appears to be approximately plane, and far from its source a spherical wave behaves very much as a plane wave would. It has very nearly the same phase constant k that a plane wave would have. As we move away from the source, the strength of the wave decreases, but far from the source the decrease in strength is inappreciable over many wavelengths.

Just how does the strength of the wave decrease with distance from the source? The power per unit area of the wave front is proportional to the square of the strength S of the wave. The total power must be independent of distance from the source. The area of the spherical surface over which the wave is spread at a distance r from the source is r^2. Hence, $S^2 r^2$ is a constant. Far from the source, the strength S of such a wave is described by

$$S = \frac{s_0}{r} e^{j(\omega t - kr)}. \tag{15.1}$$

Expression (15.1) is an *exact* solution of the linear scalar wave equation. It holds for the pressure of longitudinal waves such as sound waves. We shall see later that near the source ($r = 0$) the velocity fluctuations that accompany pressure fluctuations vary somewhat differently with distance. For some waves, such as electromagnetic waves, (15.1) accurately describes how the strengths of electric and magnetic fields vary for large values of r.

When r is very large, over a reasonable distance the wave of (15.1) behaves very nearly like a plane wave for which

$$S = S_0 e^{j(\omega t - kz)}. \tag{15.2}$$

Here z is measured in the same direction as r, and the peak strength S_0 is then given by

$$S_0 = \frac{s_0}{r}. \tag{15.3}$$

We must remember that it is only very far from sources that waves are approximately plane. Consider the situation shown in Figure 15.1. Here the source of the wave is a transmitting antenna A_c mounted on a car which is driving close to the wall of a building. A radio signal from A_c goes to a far hilltop

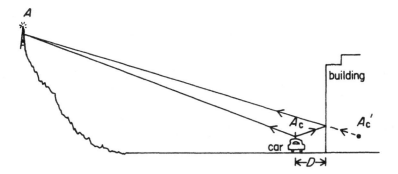

Figure 15.1

antenna A. A signal can reach A from A_c via two paths: directly, and by a reflected path. The wave that arrives via the reflected path acts as if it had come from a source $A_c{}'$ within the building, a source that is the image of the A_c seen looking at the "mirror" formed by the wall of the building. Thus, in effect, signals reach A from two sources of equal strength (for 100% reflection) but located at different distances.

The signals from the two sources may arrive at A in phase, so that the strengths add, or 180° (π radians) out of phase, so that the strengths cancel. If the strengths add at A when the car is a distance D from the wall, they cancel if we move the car a distance $\lambda/4$ closer toward or farther from the wall.

The signal received at A is very sensitive to the location of the car that transmits the signal. On the other hand, at the far hilltop the wave is approximately a plane wave, and the signal received changes only slowly as we move the hilltop antenna.

The fraction of the transmitted power that is received is the same whether we use the car antenna A_c to transmit and the hilltop antenna A to receive, or whether we use the hilltop antenna to transmit and the car antenna to receive. Hence, if we transmit from the hilltop to the car, the received power varies rapidly with the location of the car but slowly with the location of the hilltop antenna.

This different behavior with respect to location of the car antenna and location of the hilltop antenna has important consequences in mobile radio service. If we provide several antennas on a car, located at distances of about $\lambda/4$ from each other, then when nearby buildings cause one antenna to receive a very small signal, some other antenna is almost sure to receive a strong signal. Thus, combining or choosing among the signals received by the several antennas on a car helps to avoid the bad effects of reflections from nearby buildings. Such diversity can help in receiving signals on a remote hilltop, but the receiving antennas must be far apart, for the signal that reaches the hilltop from the car is nearly a plane wave, and its peak strength changes only slowly with distance.

The foregoing discussion of mobile radio shows the practical consequences of the different nature of waves in a region near to a number of sources and in a region far from a number of sources. We can illustrate this dramatically by assuming two equal and opposite sources whose strengths vary exactly in accord with (15.1). We assume the sources to be located at $z = +a/2$ and $z = -a/2$, as shown in Figure 15.2. The distance from a point P to the origin is called r, and the line from the origin to point P makes an angle θ with the z axis.

The distance L_+ between the source at $z = +a/2$ and the point P is

$$L_+ = [r^2\sin^2\theta + (r\cos\theta - a/2)^2]^{1/2}, \tag{15.4}$$

$$= r[1 - (a/r)\cos\theta + (a/2r)^2]^{1/2}. \tag{15.5}$$

The distance L_- of the source at $-a/2$ to the point P is

$$L_- = r[1 + (a/r)\cos\theta + (a/2r)^2]^{1/2}. \tag{15.6}$$

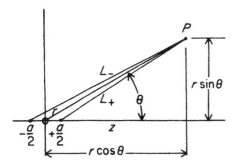

Figure 15.2

Now imagine that (15.1) holds exactly and that the source to the left is equal and opposite to the source to the right. The strengths of the two waves are not exactly equal and opposite at the point P for two reasons: (1) the right-hand source is nearer to P and hence the wave from this source has a slightly greater strength than the wave from the source to the left; (2) because the two sources are at slightly different distances, the waves from them are not exactly 180° out of phase at P.

It is simplest to investigate the consequences of these two matters separately. Let us first consider the direct effect of the difference in distance from the two sources. The series expansion for the square root of one plus a small quantity x is

$$\sqrt{1 \pm x} = 1 \pm \tfrac{1}{2}x + \ldots . \tag{15.7}$$

If we use this in connection with (15.5) and (15.6), we see that for very large values of r, very nearly

$$L_+ = r[1 - (a/2r)\cos \theta]$$
$$= r - (a/2)\cos \theta, \tag{15.8}$$

$$L_- = r[1 + (a/2r)\cos \theta]$$
$$= r + (a/2)\cos \theta. \tag{15.9}$$

If we take into account only this difference in distance in computing the distant strength due to two equal and opposite sources, we obtain as the sum of the effects of the two sources,

$$\frac{1}{L_+} - \frac{1}{L_-} = \frac{1}{r}\left(\frac{1}{1 - (a/2r)\cos\theta} - \frac{1}{1 + (a/2r)\cos\theta}\right)$$
$$= \frac{a}{r^2}\left(\frac{\cos\theta}{1 - (a/2r)^2\cos^2\theta}\right). \tag{15.10}$$

At large distances, (15.10) decreases as $1/r^2$.

Let us now examine the effect of the difference in phase of the waves of essentially equal peak strength which arrive at P from $z = +a/2$ and $z = -a/2$. We omit the common factor $e^{j\omega t}$ and assume that the strengths of the two sources are $+s_0$ and $-s_0$. We assume that (15.1) holds exactly. As to the distant strength of each source separately, we make the strengths equal by letting the factor that multiplies the exponential in (15.1) be simply s_0/r. Thus, from (15.8) and (15.9), we take the total strength S at the point P to be

$$S = \frac{s_0}{r}(e^{-jk[r - (a/2)\cos\theta]} - e^{-jk[r + (a/2)\cos\theta]});$$
$$S = \frac{2js_0}{r}(e^{-jkr})\sin[(ka/2)\cos\theta]. \tag{15.11}$$

We see that according to (15.1), at large distances the peak strength varies as $1/r$. Equation (15.10) told us that at large distances the fact that the $+$ and $-$ sources are at slightly different distances accounts for a strength that varies as $1/r^2$. Hence, we conclude that the fact that the waves from the two sources have slightly different phases, an effect expressed by (15.11), predominates over the fact that the strengths of the waves from the two sources have slightly different magnitudes, an effect expressed by (15.10). Thus, at large distances (15.11) expresses the strength of the wave from the pair of sources.

Let us consider (15.11). We now assume that $ka/2$ is very small compared with unity, that is, that the sources are separated by a very small fraction of a wavelength. The sine of a very small angle is nearly equal to the angle, so that when $ka/2$ is very small we can write S as

$$S = 2(\frac{j}{r})(e^{-jkr})(s_0 ka)\cos \theta. \tag{15.12}$$

Here S is the strength of a wave a great distance from a *dipole source*. At a great distance from the source, the wave is very nearly a plane wave. The strength of the wave varies with distance r from the source and angle θ with respect to the axis through the two sources that constitute the dipole. The strength of the wave is proportional to $s_0 ka$. This can also be written

$$s_0 ka = 2\pi s_0 a/\lambda. \tag{15.13}$$

Suppose that we assume that the sources placed at $+a/2$ and $-a/2$ radiate directionally, so that the strength of the wave from each source varies with θ and r as

$$\frac{AF(\theta)}{r},$$

where A is some multiplying factor. We can write this into (15.12) by replacing s_0 by $AF(\theta)$:

$$S = 2(\frac{j}{r})(e^{-jkr})(ka)[AF(\theta)\cos \theta]. \tag{15.14}$$

Now let us designate by S_1 the strength for a dipole made up of two *isotropic* sources for one of which

$$AF(\theta) = s_0.$$

The other source is -1 times this. Then

$$S_1 = 2(\frac{j}{r})(e^{-jkr})s_0 ka \cos \theta. \tag{15.15}$$

We can regard such a dipole as a source for which

$$AF(\theta) = 2js_0 ka \cos \theta. \tag{15.16}$$

For a pair of such sources, positive and negative, centered at $+a/2$ and $-a/2$, (15.14) says that the strength of the distant field, which we shall call S_2, is

$$S_2 = 2(\frac{j}{r})(e^{-jkr})(ka)(2js_0 ka \cos \theta)\cos \theta,$$
$$= 2^2(\frac{-1}{r})(e^{-jkr})s_0(ka)^2 \cos^2\theta. \tag{15.17}$$

Again we can regard this as a source for which

$$AF(\theta) = -2^2 s_0(ka)^2 \cos^2\theta. \tag{15.18}$$

For a pair of such sources, positive and negative, centered on $+a/2$ and $-a/2$, (15.14) tells us that the distant field strength S_3 is

$$S_3 = 2^3(\frac{-j}{r})(e^{-jkr})s_0(ka)^3 \cos^3 \theta. \tag{15.19}$$

We can easily generalize through n steps and say

$$S_n = \frac{(2j)^n s_0}{r} e^{-jkr}(ka)^n \cos^n\theta. \tag{15.20}$$

The radiation patterns S_1, S_2, S_3 correspond to increasingly complicated arrays of simple sources, each of a strength that is an integral multiple of s_0. Then S_1 corresponds to a dipole consisting of two isotropic sources at $+a/2$ and $-a/2$, and S_2 corresponds to two dipoles centered at $+a/2$ and $-a/2$. If we actually add up the isotropic sources, some of which coincide in location, we get increasingly complicated arrays of isotropic sources, as shown in Figure 15.3. The sources in these arrays are spaced a distance a apart. The magnitudes of the sources of S_n

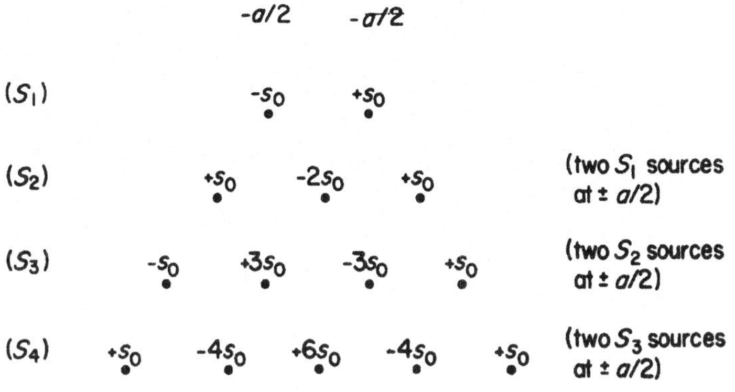

$-a/2$ $-a/2$

(S_1) $-s_0$ $+s_0$
                                  •   •

(S_2) $+s_0$ $-2s_0$ $+s_0$ (two S_1 sources
                          •   •   •   at $\pm\, a/2$)

(S_3) $-s_0$ $+3s_0$ $-3s_0$ $+s_0$ (two S_2 sources
                  •   •   •   •   at $\pm\, a/2$)

(S_4) $+s_0$ $-4s_0$ $+6s_0$ $-4s_0$ $+s_0$ (two S_3 sources
          •   •   •   •   •   at $\pm\, a/2$)

Figure 15.3

are given by the binomial coefficients of the terms in

$(-a + b)^n$.

The binomial coefficients for $a^{n-k}b^k$ are given by

$$\frac{n!\,(-1)^{(n-k)}}{(n-k)!\,k!}. \tag{15.21}$$

Here $k = 0$ corresponds to the *first* coefficient.

The power density in a direction making an angle θ with the axis is proportional to the square of the peak value of S_n, that is, to $\cos^{2n}\theta$. If we plot $\cos^{2n}\theta$ vs θ we obtain a *radiation pattern*. The distance of a point from the origin in a given direction is proportional to the power density of the wave traveling in that direction. In Figure 15.4, $\cos^{2n}\theta$ is plotted against θ for $n = 1, 2,$ and 3.

Radiation patterns according to (15.20) are symmetrical about the origin. This must be true for arrays of sources, as shown in Figure 15.3, in which the phases of the isotropic sources differ by 0° or 180°.

In a wave that traveled to the right with a phase constant k, the phase would change by e^{-jka} between a point to the left of

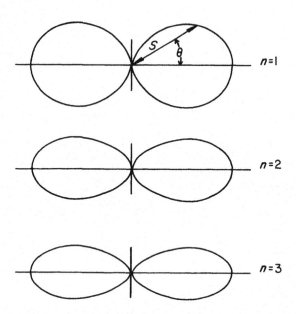

Figure 15.4

the origin at $-a/2$ and a point to the right of the origin at $+a/2$.

By making the relative phases of the sources correspond to the phase of a wave traveling in the $+z$ direction, or to this phase plus 180° (π radians), we can construct arrays of sources that radiate in the $+z$ direction and not in the $-z$ direction. We assume a negative source of phase $e^{jka/2}$ at $-a/2$ and a positive source of phase $e^{-jka/2}$ at $+a/2$. We assume the peak strength of radiation from these sources at large distances r to be

$$\frac{AF(\theta)}{r}.$$

Then for large values of r the total strength S of the wave produced by the two sources is

$$S = \frac{AF(\theta)}{r}(e^{-jk[r-(a/2)\cos\theta-(a/2)]}$$
$$- e^{-jk[r+(a/2)\cos\theta+(a/2)]})$$

$$= \frac{2jAF(\theta)}{r} e^{-jkr} \sin[(ka/2)(1 + \cos \theta)]. \tag{15.22}$$

As before, we assume that $ka/2$ is very small so that we can take the sine of the angle as equal to the angle. We also note that

$$1 + \cos \theta = 2 \cos^2(\theta/2). \tag{15.23}$$

We thus obtain

$$S = \frac{2jA}{r} e^{-jkr}(ka)F(\theta)\cos^2(\theta/2). \tag{15.24}$$

We call the strength S of (15.24) S_1 when

$$AF(\theta) = s_0. \tag{15.25}$$

Thus,

$$S_1 = \frac{2js_0}{r} e^{-jkr}(ka)\cos^2(\theta/2). \tag{15.26}$$

We call S of (15.24) S_2 when

$$AF(\theta) = 2js_0(ka)\cos^2(\theta/2). \tag{15.27}$$

Thus

$$S_2 = \frac{-4s_0}{r} e^{-jkr}(ka)^2 \cos^4(\theta/2). \tag{15.28}$$

In general, we see that

$$S_n = \frac{(2j)^n s_0}{r} e^{-jkr}(ka)^n \cos^{2n}(\theta/2). \tag{15.29}$$

The distribution of sources that produce the radiation patterns given by (15.29) are the same as those of Figure 15.3 except for an added phase e^{-jka} at each source. The power density radiated in a given direction is proportional to the square of the peak strength and hence varies with θ as $\cos^{4n}(\theta/2)$. In Figure 15.5, $\cos^{4n}(\theta/2)$ is plotted against θ for $n = 1, 2,$ and 3.

The radiation patterns we have discussed result from arrays of isotropic sources. Longitudinal waves, such as sound waves, can be radiated isotropically. There are no isotropic sources for electromagnetic waves and for other transverse waves that exhibit polarization. For such waves, the simplest source is a dipole source.

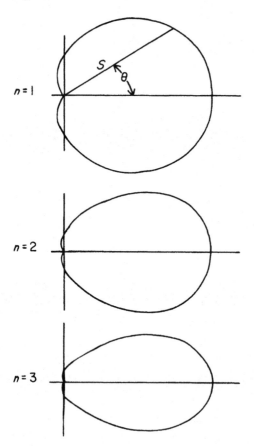

Figure 15.5

Dipole sources of electromagnetic waves are of two kinds, as shown in Figure 15.6. In a we have a short wire with a sinusoidal current flowing in it in the y direction, which is perpendicular to the z direction. In b we have a small loop of wire in which a sinusoidal current flows. The loop circles the y axis, and lies in a plane perpendicular to the y axis.

In a direction making an angle ψ with the y axis, far from the origin, the strength S of the wave from such a source can be expressed as

$$S = \frac{s_1}{r} e^{-jkr} \sin \psi. \tag{15.30}$$

Electromagnetic waves have an electric field and a magnetic field. Both fields are perpendicular to the direction of propagation (the direction in which r lies) and the electric field is perpendicular to the magnetic field. For the short wire in a of Figure 15.6, the electric field lies in a plane through the y axis. For the loop of b of Figure 15.6, the magnetic field lies in a plane through the y axis.

Expressions (15.20) and (15.29) can be modified to apply to radiation from such dipole sources by replacing s_0 by $s_1 \sin \psi$. They are then correct for large values of r.

A few of the radiation patterns we have discussed actually occur in practical applications. A device that produces a given radiation pattern when it is supplied with energy exhibits the same directional characteristics in picking energy up.

As an example, some microphones respond to the pressure of a sound wave. If they are driven electrically, they radiate equally in all directions; they act as isotropic sources. They also are equally sensitive to plane waves reaching them from any direction. A microphone that responds to pressure only is nondirectional.

Other microphones (*ribbon microphones*) respond to the veloc-

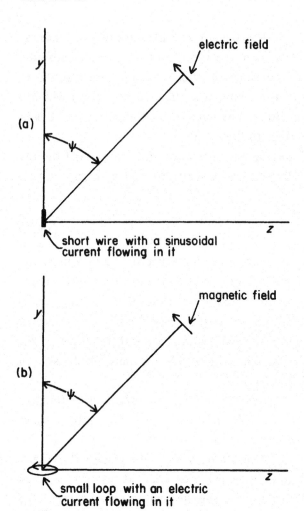

electric field

(a)

ψ

y

z

short wire with a sinusoidal
current flowing in it

magnetic field

(b)

ψ

y

z

small loop with an electric
current flowing in it

Figure 15.6

ity of the sound wave. If driven electrically, they produce a dipole radiation pattern, according to (15.15). They also respond to plane waves reaching them from various directions in accord with the same angular pattern.

Some microphones have the *cardioid* pattern of (15.26). This is useful in picking up the speaker's voice while avoiding singing due to sound from loudspeakers in the auditorium. The cardioid pattern is usually obtained by combining a pressure-sensitive microphone with a velocity-sensitive microphone; for the combination the phases of the waves radiated are in phase in one direction and out of phase in the opposite direction.

Microphones with more complicated directional patterns are sometimes used. When energy is fed to such microphones, they produce strong sound pressures and velocities nearby but weak distant fields. As microphones, they respond to nearby sources, which produce waves whose wavefront in the vicinity of the microphone is appreciably curved, but they have little response to the (nearly) plane waves from distant sources.

So far we have considered the field that an array of sources produces at a great distance, where the wave produced is close to a plane wave. We have seen that the angular variation of radiation according to (15.20) and (15.29) does not depend at all on the size of the array of sources, that is, on the distance a between the sources. It is theoretically possible to radiate an extremely narrow beam of waves (by making n large) from sources in as small a space as we wish (by making a small). This was not appreciated in the early days of radio; it was believed that a large antenna was required in order to produce a narrow beam of waves. We shall see why in Chapter 16. Prior to 1943, S. A. Schelkunoff[1] realized that this was not so, and he devised means for producing small *superdirective* antennas. Since then, the possibility of superdirective antennas has been rediscovered by a number of people.[2] The reader may wonder why such antennas are not used.

By examining Figure 15.3 we see that the greater the directivity (the greater n), the greater the strengths and number of the sources that are needed. By examining (15.20) and (15.29) we see that the greater the directivity, the smaller is the field *at large values of r*, because the expression for the strength has $(ka)^n$ as a factor, and ka is assumed to be very small. Thus, the smaller a is, the smaller is the strength of the wave at great distances compared with *both* the strength of the sources (an integer times s_0) and the strength of the waves *near* the sources, where (15.20) and (15.29) do not hold because $ka/2$ is not small compared with unity.

As we make an antenna (an array of sources) which has a given directional pattern smaller and smaller, the strength of the waves near the antenna becomes larger and larger compared to the strength of the waves at a great distance. This means that for a given radiated power the electric currents in the antenna become larger and larger as the antenna is made smaller and smaller.

There are large power losses in conductors in which large currents flow, so superdirective antennas become less and less efficient as they are made smaller and smaller. Further, in a superdirective antenna the required phase relations among the sources can be maintained over a narrow range of frequencies only. Superdirective antennas are not of practical use. However, we shall find other uses for the *ideas* that we have explored in this section.

References

1. S. A. Schelkunoff, U. S. Patent 2,286,839; also, A Mathematical Theory of Linear Arrays, *Bell System Technical Journal*, Vol. 22, pp. 80–107 (January 1943).

2. L. J. Chu, Physical Limitations of Omni-Directional Antennas, *Journal of Applied Physics*, Vol. 19, pp. 1163–1175 (December 1948).

Problems

1. In Figure 15.4 (Equation 15.20), what is the slope of the curve at the origin ($\theta = \pi/2$)?

2. In Figure 15.5 (Equation 15.29), what is the slope of the curve at origin ($\theta = \pi$)?

3. Assume a *continuous* distribution of isotropic sources along the z axis, extending from $z = -L/2$ to $z = +L/2$. Consider the wave strength S at a point a long distance r from the origin ($z = 0$). Let the line between the point and the origin make an angle θ with the z axis. Let the portion of the strength at the distant point dS due to sources in the distance dz at z be

$$dS = \frac{S_0}{LD} e^{-jkD} dz.$$

Here D is the distance from the point z on the z axis to the distant point. What is the total field S as a function of r and θ? Are there values of L for which the pattern is simpler than for other values? In radio, such a distribution of sources is called a *broadside array*. Why?

4. Rework the problem as stated in 3 except assume that

$$dS = \frac{S_0}{LD} e^{-jkD} e^{-jkz} dz.$$

In radio such a distribution of sources is called an *endfire array*. Why?

16 Antennas and Diffraction

Figure 16.1 represents a beam of light emerging from a laser. As the beam travels, it widens and the surfaces of constant phase become spherical. The beam then passes through a convex lens made of a material in which light travels more slowly than in air. It takes a longer time for the waves to go through the center of the lens than through the edge of the lens. The effect of the lens is to produce a plane wave over the area of the lens. When the light emerges from the lens, the wavefront, or surface of constant phase, is plane.

Figure 16.2 represents one type of microwave antenna. A microwave source, such as the end of a waveguide, is located at the focus of a parabolic (really, a paraboloidal) reflector. After reflection, the phase front of the wave is plane over the aperture of the reflector.

The light emerging from the lens of Figure 16.1 does not travel forever in a beam with the diameter of the lens. The microwaves from the parabolic reflector of Figure 16.2 do not travel forever in a beam the diameter of the reflector. How strong is the wave at a great distance from the lens or the reflector?

A particular form of this question is posed in Figure 16.3. We feed a power P_T into an antenna that emits a plane wave over an area A_T. We have another antenna a distance L away which picks up the power of a plane wave in an area A_R and supplies this power P_R to a receiver. What is the relation among P_T, P_R, A_T, A_R, and L? In 1946 Harald T. Friis gave a very simple formula relating these quantities:

$$\frac{P_R}{P_T} = \frac{A_R A_T}{\lambda^2 L^2}. \tag{16.1}$$

Here λ is the wavelength. Equation 16.1 holds if the transmitting

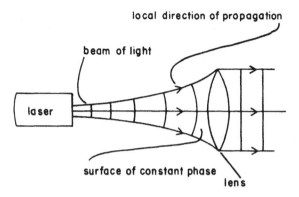

Figure 16.1

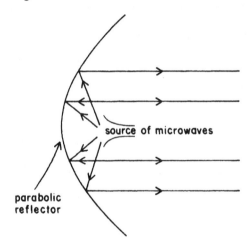

Figure 16.2

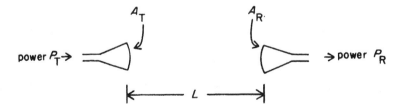

Figure 16.3

antenna produces a wave of constant strength over the aperture of area A_T and if the receiving antenna picks up all the energy of a plane wave over the area A_R.

When an antenna is used as a transmitting antenna, it may produce a wave whose strength varies over the aperture. In this case it has an *effective area*, which is less than its geometrical area, and this effective area should be used in (16.1). The effective area of an antenna when used as a transmitting antenna is the same as its effective area when used as a receiving antenna.

Friis's transmission formula is so simple that one feels that there must be a simple way of deriving it. It is easy to see that P_R / P_T must be given by a constant times $A_R A_T / \lambda^2 L^2$, but it is not easy to see that the constant is unity. Friis published the first simple derivation in 1971.[1] He made use of an interesting procedure that derives from Christian Huygens. In his *Traité de la lumière*, published in 1690, Huygens stated that every point on a wavefront is the center of a new wave. This idea was effectively put to use by Fresnel in his *Memoir on the Diffraction of Light*, published in 1816.

The word *diffraction* is used to describe the representation of a wave and calculations concerning waves in which we replace the wavefront by sources of new waves. What are we to make of Huygens' assertion that every point on a wavefront is the center of a new wave? What sort of source can replace and represent a wavefront exactly?

Figure 16.4 represents a plane wave traveling to the right. Suppose that within a narrow width w we put something that absorbs the wave traveling from the left and something that generates an equal wave to the right. Outside the space of width w, nothing is different from what it was before.

What can we put within the width w to accomplish this?

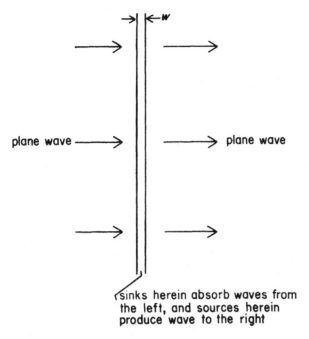

Figure 16.4

Equation 15.29 describes the radiation of an array of point sources which sends no wave to the left ($\theta = \pi$). If we distribute the point sources over a plane normal to the z axis, we form an array of plane sources which sends a plane wave to the right but not to the left.

In deriving (15.29) we made use of isotropic *sources* which produce waves that vary with r as

$$\frac{e^{-jkr}}{r}.$$

We can also have isotropic *sinks* that absorb waves that vary with r as

$$\frac{e^{+jkr}}{r}.$$

We can construct an array of plane sinks which just absorbs all of the wave arriving from the left, in the same way that we arrived at (15.29). In fact, we need only replace e^{-jkr} by e^{jkr} to go from sources to sinks, and phasing the sinks correctly takes us from $\cos(\theta/2)$ to $\cos(\pi + \theta/2)$.

We are left with the choice of n. We can choose any value of n. If we take the array of point sources corresponding to *any* value of n and smear them uniformly over planes normal to the z axis, we get an array of plane sources which produce a plane wave to the right and no wave to the left!

There are many valid ways of representing a plane wave as the sum of waves radiated from every point on the wavefront. We can regard a plane wave as a sum of narrow radiated beams (n large) or of broad radiated beams (n small). One representation is as valid as another.

In fact, to the right of the volume of width w in Figure 16.4, we cannot tell what happens to the left. Hence, if we are interested in waves to the right of the wavefront only, we can replace the wavefront by any sources that add up to give a plane wave to the right, even if these sources give the wrong wave to the left. We can represent the wave to the right by any of the field patterns of (15.20), which radiate to the left as well as to the right. All of these give equally valid representations of the plane wave to the right of their location.

We now consider the problem shown in Figure 16.5. A plane wave comes from the left, traveling in the z direction. Over the plane normal to z at the origin we replace the incoming plane wave by sources that produce the same plane wave to the right. The sources on any small area of the plane have a directional pattern $F(\theta)$ with respect to the z direction. We are free to choose any of many different radiation patterns $F(\theta)$. It is advantageous to choose any one for which $F(\theta)$ goes monoton-

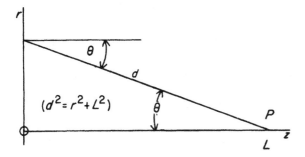

Figure 16.5

ically to zero as θ approaches $\pi/2$. The radiation patterns of (15.20) fulfill this criterion.

Let us now consider the plane normal to the z axis at $z = 0$, over which there is a uniform distribution of sources. Figure 16.5 shows this as seen normal to the z axis. The area A within a radius r of the origin is

$$A = \pi r^2 = \pi(d^2 - L^2). \tag{16.2}$$

Suppose we increase d by a small amount Δd. The area increases by a small amount ΔA given by

$$\Delta A = 2\pi d\, \Delta d. \tag{16.3}$$

The total strength of the sources in an area is proportional to the area. The strength of the field that the sources produce at P is proportional to the area divided by d and multiplied by $F(\theta)$. We then write this increment ΔS in field strength at P as

$$\Delta S = \frac{K\, \Delta A\, F(\theta)}{d} = 2\pi K F(\theta)\, \Delta d. \tag{16.4}$$

Here K is a constant.

When we increase the distance from a source by an amount Δd, we change the phase of the wave by the phase constant k

CHAPTER 16   166

times this distance. All the sources in the plane are in phase. Thus, the change in phase $\Delta\phi$ corresponding to a change of distance Δd is

$$\Delta\phi = -k\Delta d = \frac{-2\pi\Delta d}{\lambda}. \tag{16.5}$$

We can regard the field at P as made up of many small components corresponding to sources in the small areas ΔA, areas specified by *equal* increments of distance, Δd, and of phase, $\Delta\phi$. The length or magnitude ΔS of each component is given by (16.4). The phase change from one component to the next is given by (16.5). The first components lie near the z axis; for them $F(\theta) = 1$, and so all these components are equal in length. Thus they must lie around the circumference of a circle, as shown in Figure 16.6.

As we include more and more of the plane by adding little areas ΔA corresponding to the equal increases Δd in distance, the sum of the components moves around the circle shown in Figure 16.6. As we move around the circle many times, the areas we add are farther and farther from the axis. Thus, θ increases and $F(\theta)$ decreases, and the *length* of the added component decreases, but the angle between successive added components remains the same. This means that the sum of the components ultimately spirals inward and reaches the origin as $F(\theta)$ goes to zero. Thus, the magnitude of the field at P due to all the sources on the whole plane is the distance from the starting point O to the center of the circle, that is, the radius S_0 of the circle.

What is the ratio of the strength ΔS due to sources in a small circle of area ΔA about the origin, and the strength S_0 due to all the sources in the plane?

From Figure 16.6 and (16.5) we can find a relation involving the radius S_0 of the circle, the increment of field strength ΔS for $F(\theta) = 1(\theta$ small$)$, Δd, and λ:

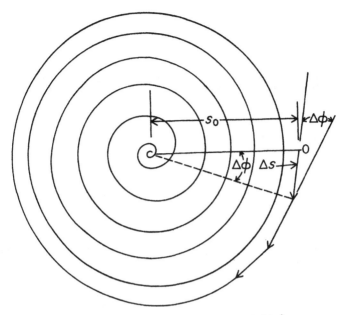

Length of component ΔS proportional to $2\pi F(\theta)\Delta\phi$.
Angle $\Delta\phi$ between components proportional to $k\Delta d$.

Figure 16.6

$$\Delta\phi = \frac{-\Delta S}{S_0} = \frac{-2\pi\,\Delta d}{\lambda}. \tag{16.6}$$

From (16.3) we see that, near the axis where $d = L$,

$$2\pi\,\Delta d = \frac{\Delta A}{L}. \tag{16.7}$$

Hence, from (16.6) and (16.7)

$$\frac{\Delta S}{S_0} = \frac{\Delta A}{\lambda L}. \tag{16.8}$$

Here ΔS is the peak strength of the field at P due to sources in an area ΔA about the origin, and S_0 is the peak strength of the field at P due to sources over the whole plane. As these sources produce a plane wave, S_0 must also be the peak field strength at the origin. Power density at the origin is proportional to S_0^2; power density at P is proportional to $(\Delta S)^2$. We let ΔA be the area of the transmitting antenna

$$\Delta A = A_T \tag{16.9}$$

The ratio of the total received power P_R to the total transmitted power P_T is

$$\frac{P_R}{P_T} = \frac{A_R}{A_T}\left(\frac{\Delta S}{S_0}\right)^2. \tag{16.10}$$

From (16.8)–(16.10) we have

$$\frac{P_R}{P_T} = \frac{A_R}{A_T}\left(\frac{A_T^2}{\lambda^2 L^2}\right) = \frac{A_R A_T}{\lambda^2 L^2}. \tag{16.11}$$

This is just Friis's transmission formula, which is used in designing microwave systems and in laser work.

Friis's transmission formula, that is (16.1) or (16.11), holds only when each antenna subtends a very small angle as seen from the other antenna, and when the antennas are pointed directly at one another. Within this restriction, it holds for either longitudinal waves such as sound waves or transverse waves such as electromagnetic waves. The restriction to antennas of small apparent diameter assures that in the case of transverse waves the strengths of the waves from all areas of the transmitting antenna have the same direction at the receiving antenna.

Now that we have derived the transmission formula, it is of some interest to discuss it.

Far from the transmitting antenna, the wave diverges radially from the source and the power P_T is spread over an area proportional to L^2. This is in accord with the term $1/L^2$. The A_R must appear as a factor because far from the source the power density changes slowly with distance, and the total power the receiving antenna picks up is proportional to its area A_R.

For a pair of antennas, if we interchange the power source and the receiver the fraction of power received is the same. Thus A_T must appear as a factor.

What about λ? Suppose that we look at a transmitting antenna from a great distance. As we move off axis, one side of the antenna is farther away from us than the other, and waves from the far side are out of phase with waves from the near side. The shorter the wavelength, the more out of phase the waves from the two edges of the antenna are. In fact, an antenna of width W emits a beam of radiation whose minimum possible angular width θ is proportional to λ/W. The smaller λ, the narrower the beam. At a distance, the transmitter power is spread over an area proportional to the angle times the distance L, and the power density at the receiving antenna is inversely proportional to this area. That is, the power density at the receiving antenna is proportional to

$$\frac{P_T W^2}{\lambda^2 L^2}. \tag{16.12}$$

In (16.1) W^2 is represented by the area A_T, and P_T, L^2, and λ^2 occur as expected.

The sort of argument we have given above gives the necessary form of the transmission formula, but the derivation given in Equations 16.2–16.11 gives the exact numerical relation.

The transmission formula says that the received power is proportional to the area of the transmitting antenna. However, in the derivation we have assumed that the transmitting antenna is small. What we *can* say is that when the transmitting antenna is small the power density at the receiving antenna, P_R/A_R, must be related to the power density at the transmitting antenna, P_T/A_T, by

$$\frac{(P_R/A_R)}{(P_T/A_T)} = \frac{A_T^2}{\lambda^2 L^2}. \tag{16.13}$$

This means that the strengths of the waves at the receiving and transmitting antennas, S_R and S_T, must be related by

$$\frac{S_R}{S_T} = \frac{A_T}{\lambda L}. \tag{16.14}$$

This holds for small areas A_T only.

Let us now imagine that sources of uniform strength are uniformly distributed over a spherical cap of large radius L and of half-angle α, as shown in Figure 16.7. The area A of such a cap is

$$\begin{aligned}
A &= \int_0^\alpha (2\pi L \sin \alpha) L \, d\alpha \\
&= 2\pi L^2 (1 - \cos \alpha) \\
&= 4\pi L^2 \sin^2(\alpha/2).
\end{aligned} \tag{16.15}$$

If sources are distributed uniformly over the surface, the *strengths* of their waves all *add* at the center and are proportional to A/L. Thus, we can write the field strength S_R at the center,

$$S_R = C \sin^2(\alpha/2). \tag{16.16}$$

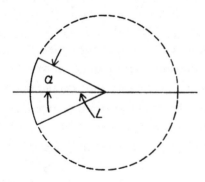

Figure 16.7

Here C is a multiplying factor which we must somehow determine.

When α is very small, the spherical cap becomes indistinguishable from a very small plane disk. We can apply (16.14) to this case. When α is very small,

$$A_T = \pi \alpha^2 L^2. \tag{16.17}$$

Thus, for very small values of α we have from (16.14)

$$\frac{S_R}{S_T} = \frac{\pi \alpha^2 L}{\lambda}. \tag{16.18}$$

From (16.16) we have for very small values of α

$$S_R = \frac{C\alpha^2}{4}. \tag{16.19}$$

We see that

$$C = \frac{4\pi S_T L}{\lambda}, \tag{16.20}$$

and, from (16.16)

$$\frac{S_R}{S_T} = 4\pi \left(\frac{L}{\lambda}\right)\sin^2(\alpha/2). \tag{16.21}$$

We might have naïvely thought of a wave as coming in toward the origin and becoming infinite there. The wave just doesn't do this. It is diffuse around the origin. It crowds itself into a region whose size is proportional to the wavelength divided by $\sin^2(\alpha/2)$, and then it spreads out again.

Suppose that the wave power remained constant within the cone of half-angle α as the wave traveled inward. Then the strength ratio of (16.21) would occur a distance $\lambda/4\pi \sin^2(\alpha/2)$ from the origin. At this distance the half-width of the beam, a, is

$$a = [\lambda/4\pi \sin^2(\alpha/2)]\sin \alpha. \tag{16.22}$$

Actually, the wave travels somewhat as shown by the dashed curves of Figure 16.8, and a affords some estimate of the beam radius at the origin. Of course, part of the beam is inside of this radius, and part outside. For small values of α, (16.22) becomes

$$a = \lambda/\pi\alpha. \tag{16.23}$$

Equation 16.21 holds for polarized waves, such as electromagnetic waves, for small values of α only. At larger angles S_R is smaller than given by (16.21), because the strengths of the waves arriving at the origin from various parts of the spherical cap have different directions, and also because of the nonisotropic nature of sources of polarized waves.

In this chapter we have replaced the whole of a wavefront by sources and have then calculated the distant field of these sources. In some diffraction calculations, one is concerned with a wave that falls on an opaque plate that is pierced by two or more small apertures. Each aperture acts as a source and one computes the distant field as the sum of the waves from these sources.

In such a case, the distant field depends in part on the directional pattern of the wave from a single aperture. The wave that comes out of a small aperture is not merely a sample of the wave that struck the apertured plate. The wave interacts with the apertured plate. It is only by a detailed calculation that we can find the nature of the source which adequately represents the wave emerging from the aperture.

Insofar as the directional pattern of the waves from the apertures is not important, we can get valid results by merely considering the relative phases of the waves from various apertures. Thus, we can calculate correctly the relative strengths of waves from the apertures over any region in which all apertures are seen in approximately the same direction. If we

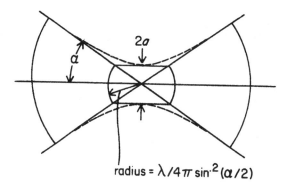

radius = $\lambda/4\pi \sin^{-2}(\alpha/2)$

Figure 16.8

want to compute actual field strengths, we must know the actual strengths of the sources, and if we want to know strengths over a wide range of angles, we must know the directional pattern of the sources.[2]

References

1. H. T. Friis, Introduction to Radio and Antennas, *IEEE Spectrum*, Vol. 8, pp. 55–56 (April 1971).

2. Sergei A. Schelkunoff and Harald T. Friis, *Antennas, Theory and Practice*, New York: John Wiley & Sons, 1952.

Problems

1. In the derivation of Equation 16.11 it is assumed that the strength of the field is constant over the geometrical aperture or area of the transmitting antenna. Either antenna can be used as the transmitting antenna, so that it has been assumed that if either antenna is used as a transmitting antenna, the strength of the field will be constant over the aperture. In actual antennas, the strength of the field varies over the aperture. In this case (16.11) holds if A_T and A_R are interpreted as *effective* areas

which differ from the geometrical area of the aperture of the antenna. What is the effective area of an antenna if the field strength varies as $S = F(r)$ from $r = 0$ to $r = a$ and is zero beyond this radius?

2. Can you give a more general result than that of problem 1?

3. An interferometer consists of two antennas that point in the same direction but are separated by many wavelengths. Give an expression for the directional pattern of an interferometer in terms of the directional patterns and the separation of the two antennas.

4. We will represent a square microwave antenna by sources which are uniformly distributed over a square area normal to the z axis and centered on the origin. The portion of the strength at a point a great distance D from the point $x,y,0$ due to sources at $x,y,0$ will be

$$dS = \frac{S_0 F(\theta)}{L^2 D} e^{-jkD} dx\, dy.$$

Here L is the length of the side of the square and θ is the angle between the z axis and the line D, which extends from $x,y,0$ to the distant point. We assume that θ is small and $F(\theta) = 1$. For D very large, integrate to obtain the distant field in terms of the coordinates of the distant point. Express the result in terms of the distance of the distant point from the origin and angles with respect to the z axis.

17 The Scalar Wave Equation

Waves that travel in the z direction and whose strength does not vary with x or y obey the following equation:

$$\frac{\partial^2 S}{\partial z^2} = \frac{1}{v^2}\frac{\partial^2 S}{\partial t^2}. \tag{17.1}$$

A general solution of (17.1) is

$$S = F(t \pm z/v). \tag{17.2}$$

That is, *any* function of the quantity $(t \pm z/v)$ is a solution of (17.1). Thus, S can vary in any way with distance (at a given time), and this pattern can travel to the left or to the right with a velocity v. Equation 17.1 applies to perfectly plane electromagnetic waves, and it almost exactly describes plane sound waves and waves in stretched strings.

The wave equation corresponding to (17.1) which holds for three dimensions is

$$\frac{\partial^2 S}{\partial x^2} + \frac{\partial^2 S}{\partial y^2} + \frac{\partial^2 S}{\partial z^2} = \frac{1}{v^2}\frac{\partial^2 S}{\partial t^2}. \tag{17.3}$$

We see that this reduces to (17.1) if S does not change in the x and y directions, that is, if we have a truly plane wave.

We have noted that (17.1) describes a plane wave. Two solutions of (17.1) are

$$S_R = S_0 e^{j[\omega t - (\omega/v)z]}, \tag{17.4}$$

$$S_L = S_0 e^{j[\omega t + (\omega/v)z]}; \tag{17.5}$$

S_R is a plane wave traveling to the right; and S_L is a plane wave traveling to the left.

We can combine the waves of (17.4) and (17.5) to give a wave

S_1, where

$$S_1 = (S_L - S_R)/2j. \tag{17.6}$$

We see that

$$S_1 = S_0 e^{j\omega t} \sin[(\omega/v)z]. \tag{17.7}$$

Equation (17.7) describes a standing wave. The peak strength varies with z as $\sin[(\omega/v)z]$; at each point the strength varies sinusoidally with time. We get a wave such as that described by (17.7) when a plane wave is reflected by a perfectly reflecting surface which is normal to the direction in which the wave travels.

We now turn from plane waves to spherical waves. If S varies only with distance r from the origin (spherical waves), we can write

$$\frac{\partial^2 S}{\partial r^2} + \frac{2}{r}\frac{\partial S}{\partial r} = \frac{1}{v^2}\frac{\partial^2 S}{\partial t^2}. \tag{17.8}$$

An exact solution of (17.8) is

$$S_1 = \frac{s_0}{r} e^{j[\omega t - (\omega/v)r]}, \tag{17.9}$$

$$S_2 = \frac{s_0}{r} e^{j[\omega t + (\omega/v)r]}; \tag{17.10}$$

S_1 represents a wave traveling outward from a source of zero dimensions, and S_2 represents a wave traveling inward to a sink of zero dimensions.

In Chapter 15 we assumed a wave whose strength varied with r according to (17.9). We also assumed that the power of the wave varied as $1/r^2$. All of this is true, and yet we can easily miss an important aspect of spherical waves if we say no more. Let us recall something we learned about conical horns in Chapter 9.

We found the equation for a wave traveling in a conical horn to be

$$\frac{d^2 V}{dz^2} + \frac{2}{z}\frac{dV}{dz} + k_1^2 V = 0. \tag{17.11}$$

In (17.8) let S vary as $e^{j\omega t}$, and let

$$k = \frac{\omega}{v}. \tag{17.12}$$

Then (17.8) becomes

$$\frac{\partial^2 S}{\partial r^2} + \frac{2}{r}\frac{\partial S}{\partial r} + k^2 S = 0. \tag{17.13}$$

This is of exactly the same form as (17.11). Indeed, if we let the horn to which (17.11) applies open up to an aperture of 360° or 2π radians, as shown at the right in Figure 9.1, the horn disappears, and we have merely a spherical wave, to which Equations 17.8 and 17.13 apply.

In Chapter 9 we found that there were two variables, V and W, associated with the wave in the horn. One of these, V, varied with distance according to (17.9) or (17.10). The strength fell off simply as $1/r$. The other had a more complicated variation with distance. The complex power, V^*W was

$$V^*W = (V_0^2/A)(\mp 1 - j/k_1 z). \tag{9.42}$$

The complex power of the waves described by (17.9) and (17.10) must have the same variation with distance as that given by (9.42), because, as we have seen, the differential equations (17.11) and (17.8), the equations that govern the waves, are the same.

We can measure the strengths S_1 and S_2 in any units we choose. If we make the real power unity, then we see from (9.42) that the complex power P_c must be

$$P_c = (1 - j/kr).$$ (17.14)

We see that near the origin of a spherical wave there must be large field strengths such that energy is alternately pumped out into the space around the origin and then drawn back into the source.

In Chapter 16 we saw that there must be very large fields near small directional sources which are made up of closely spaced positive and negative isotropic sources. Now we see that there must be large fields near an isotropic source itself.

This is a general characteristic of sources. The power from a source that is very small compared with a wavelength has a very large reactive component. The physical nature of this reactive component depends on the physical nature of the wave, but the reactive component is necessarily there.

A sound wave involves both pressure and velocity. The pressure per unit area in a spherical sound wave varies in accordance with (9.40). Let us call this pressure V and the distance from the origin r; it thus varies as

$$V = \frac{V_0}{r} e^{\pm kr}.$$ (17.15)

If W is the velocity associated with the pressure V (*not* the wave velocity), W varies in accordance with (9.41), as

$$W = \frac{W_0}{r}\left(\mp 1 - \frac{j}{kr}\right)e^{\pm jk_1 r}.$$ (17.16)

The complex power is $4\pi r^2$ times V^*W; we see that the variation with distance is in accord with (9.42).

In such a sound wave, near the origin there is a great deal of kinetic energy associated with matter moving rapidly near the sound source. This is so for small sources of finite size. For example, we can imagine an underwater sound source in the form of a spherical balloon whose internal air pressure varies sinusoidally with time (because of an air pump). When k times the radius of the balloon is small, the chief force that the water exerts on the balloon is very nearly that necessary to keep a constant mass (of water) in sinusoidal motion. Very little power is radiated. Chiefly, kinetic energy is alternately imparted to and withdrawn from the water surrounding the source.

An electromagnetic wave is radiated when a sinusoidally varying electric current flows through a loop of wire. If the loop is small compared with a wavelength, there is a large magnetic field near the loop and energy is repeatedly supplied to and withdrawn from this magnetic field as the current and the field alternately increase and decrease.

Let us now return to (17.9) and (17.10). Clearly, we do not encounter in nature sound waves whose strength goes to infinity at a physical point, as is the case for S_1 and S_2. Indeed, sound waves do *not* obey (17.8) at sources of sinks of zero size at which, according to (17.9) and (17.10), the pressure would be infinite.

However, (17.9) and (17.10) are not useless. They do satisfactorily represent spherical sound waves far from sources or sinks. And, we can combine these waves in a very interesting form. We define a wave of strength S by

$$S = (S_2 - S_1)/2j(\omega/v). \tag{17.17}$$

From (17.9) and (17.10) we see that

$$S = s_0 e^{j\omega t}\left(\frac{\sin[(\omega/v)r]}{(\omega/v)r}\right). \tag{17.18}$$

Like S_1 of (17.7), S as given by (17.18) is a *standing wave*. We can imagine that a sinsoidal spherical wave comes in toward the origin, goes right past it and comes out again.

We note that S does not go to infinity at the origin. Rather, S goes to S_0 at the origin, for when r is very small $\sin[(\omega/v)r]$ is equal to $(\omega/v)r$.

In (16.16) we have the field S_R at the origin as obtained by a diffraction calculation:

$$S_R = C \sin^2(\alpha/2). \tag{17.19}$$

When $\alpha = \pi$, the source surrounds the origin. This corresponds to a solution of the wave equation (17.18). We see from (17.18) that the ratio of the peak strength at a great distance to the peak strength at the origin is

$$\frac{1}{(\omega/v)r}. \tag{17.20}$$

However, half of the peak strength at a distance belongs to the outgoing wave. If S_T is the peak strength of the incoming wave at large distance L and S_R is the total peak strength at the origin, we see that

$$\begin{aligned}\frac{S_R}{S_T} &= 2(\omega/v)L \sin^2(\alpha/2) \\ &= \frac{4\pi L}{\lambda} \sin^2(\alpha/2).\end{aligned} \tag{17.21}$$

This is exactly the same as (16.21), which we obtained without direct recourse to the wave equation.

We notice from (17.18) that S is zero at a sphere whose radius is such that

$$\frac{\omega}{v}r = \frac{2\pi r}{\lambda} = \pi. \tag{17.22}$$

The radius of this region, r, is thus

$$r = \lambda/2. \tag{17.23}$$

We see that for $\alpha = \pi/2$, at least, (16.23) underestimates the size of the region over which the wave extends near the origin.

We have seen that (17.7) represents a standing wave in which a wave approaches the origin and, passing through or reflected by it, travels outward. We can, however, find waves that approach the origin from one side and travel out on the other. Such waves are not spherically symmetrical and hence they cannot be solutions of (17.8), but they must be solutions of (17.3), of which (17.8) is a special case.

Let us rewrite (17.3) in the following form:

$$\frac{\partial^2 S}{\partial x^2} + \frac{\partial^2 S}{\partial y^2} + \frac{\partial^2 S}{\partial z^2} - \frac{1}{v^2}\frac{\partial^2 S}{\partial t^2} = 0. \tag{17.24}$$

Note that is S is a solution of (17.24), then $\partial S/\partial z$ (or $\partial S/\partial x$ or $\partial S/\partial y$) is also a solution. We can see this in the following way:

$$\frac{\partial^2(\partial S/\partial z)}{\partial x^2} + \frac{\partial^2(\partial S/\partial z)}{\partial y^2} + \frac{\partial^2(\partial S/\partial z)}{\partial z^2} - \frac{1}{v^2}\frac{\partial^2(\partial S/\partial z)}{\partial t^2}$$
$$= \frac{\partial}{\partial z}\left(\frac{\partial^2 S}{\partial x^2} + \frac{\partial^2 S}{\partial y^2} + \frac{\partial^2 S}{\partial z^2} - \frac{1}{v^2}\frac{\partial^2 S}{\partial t^2}\right) = 0. \tag{17.25}$$

It is permissible to change the order of taking partial derivatives. A quantity that is always and everywhere zero cannot change with z, and hence the partial derivative with respect to z must be zero.

Because S as given by (17.18) is a solution of (17.8), it must be a solution of the more general equation (17.3). Let us rewrite (17.18) in terms of x, y, and z:

$$S = s_0 \frac{e^{j\omega t}\sin[k(x^2 + y^2 + z^2)^{1/2}]}{k(x^2 + y^2 + z^2)^{1/2}}. \tag{17.26}$$

We easily find that

$$\frac{\partial S}{\partial z} = s_0 e^{j\omega t}\left(\frac{z\cos[k(x^2 + y^2 + z^2)^{1/2}]}{x^2 + y^2 + z^2}\right.$$
$$\left. - \frac{z\sin[k(x^2 + y^2 + z^2)^{1/2}]}{k(x^2 + y^2 + z^2)^{3/2}}\right). \tag{17.27}$$

Let θ be the angle between the z axis and a line from the origin to the point x, y, z, and r is of course $(x^2 + y^2 + z^2)^{1/2}$. We see that

$$\frac{\partial S}{\partial z} = s_0 e^{j\omega t}\left(\frac{\cos kr}{r} - \frac{\sin kr}{kr^2}\right)\cos\theta. \tag{17.28}$$

We see that $\partial S/\partial z$ goes to zero at the origin. We also see that $\partial S/\partial z$ is a standing wave.

Because the wave equation is linear, the sum of any two solutions is a solution. By combining (17.18) and (17.28) we construct the following solution, which we will call S_1:

$$\begin{aligned} S_1 &= S - \frac{j}{k}\frac{\partial S}{\partial z} \\ &= s_0 e^{j\omega t}\left[\frac{\sin kr}{kr} + j\cos\theta\left(\frac{\cos kr}{kr} - \frac{\sin kr}{k^2 r^2}\right)\right]. \end{aligned} \tag{17.29}$$

Let us examine (17.29) for very large values of r, such that we can disregard terms in $1/r^2$. We shall write the sine and cosine functions out in terms of e^{jkr} and e^{-jkr}:

$$S_1 = \frac{s_0 e^{j\omega t}}{2jkr}[e^{jkr}(1 - \cos\theta) - e^{-jkr}(1 + \cos\theta)]. \tag{17.30}$$

We see that when $\theta = 0$ (the right half of the z axis) the wave travels in the $+r$, that is, the $+z$ direction and for $\theta = \pi$ (the left half of the z axis) the wave travels in the $-r$, that is, the $+z$ direction. When θ is neither zero nor π, there are unequal components in both the $+r$ and $-r$ direction.

As S_1 of (17.29) is itself a solution of (17.3), we can obtain other solutions by rewriting S_1 in terms of x, y, and z and repeatedly taking partial derivatives with respect to z. The higher the derivative, the more the wave is concentrated near to the z axis.

There is a wave equation for electromagnetic waves. It says,

$$\frac{\partial^2 S_x}{\partial^2 x} + \frac{\partial^2 S_x}{\partial^2 y} + \frac{\partial^2 S_x}{\partial^2 z} = \frac{1}{c^2}\frac{\partial^2 S_x}{\partial t^2}. \tag{17.31}$$

Here c is the velocity of light, and S_x in the x component of either the electric or the magnetic field. The equation also holds if we replace S_x by S_y, the y component of S, or by S_z, the z component of S.

From (17.24) we might think that there was a spherically electromagnetic wave for which

$$S_x = \frac{S_{x0}}{r} e^{-j(\omega/c)r}. \tag{17.32}$$

However, the field components of an electromagnetic wave must be solutions of Maxwell's equation as well as being solutions of (17.24). Equation 17.32 is incompatible with Maxwell's equations. We shall not discuss the details of electromagnetic waves here; we have already discussed which among the expressions derived earlier hold, exactly or appoximately, for electromagnetic waves.

Problems

1. Explain what happens to S of (17.1) and its derivatives when the function $F(t \pm z/v)$ is unity when the argument lies between -1 and $+1$ and is zero elsewhere.

2. Draw a picture showing successive stages of reflection of the wave problem 1 when (a) the strength of the reflected wave has the same sign as the strength of the incident wave, and (b) when the strength of the reflected wave is the negative of the incident wave.

3. Show by differentiation that (17.26) is a solution of (17.3).

4. Find $\partial S_R / \partial z$ (S_R given by 17.5). Show that $\partial S_R / \partial z$ is a solution of (17.4).

5. Express S_1 of (17.29) in terms of $x, y,$ and z, differentiate with respect to z, and express the result in terms of r and θ.

6. Suppose that we have two waves as described by (17.18), one of strength $s_0 = Ae^{j\pi/2}$ centered at $(\omega/v)z = -\pi/2$ and the other of strength $s_0 = Ae^{-j\pi/2}$ centered at $(\omega/v)z = \pi/2$ Find an simple expression for the field strength on the z axis very far from the origin. Compare this distant field with S as given by (17.30).

18 Radiation

Radiation is the process by which waves are generated. So far, we have taken this process for granted. If we connect an ac source to one end of an electrical transmission line (say, a pair of wires or coaxial conductors), as shown in Figure 18.1, we expect an electromagnetic wave to travel down the line. Similarly, if, as in Figure 18.1, we move a plunger back and forth in an air-filled tube, we expect an acoustic wave to travel down the tube.

Thus, we commonly associate the radiation of waves with oscillating sources. The vibrating cone of a loudspeaker radiates acoustic (sound) waves. The oscillating current in a radio or television transmitting antenna radiates electromagnetic waves. An oscillating electric or magnetic dipole radiates plane-polarized waves. A rotating electric or magnetic dipole radiates circularly polarized waves.

Radiation is always associated with motion, but it is not always associated with changing motion. Imagine some sort of fixed device moving along a dispersive medium. In Figure 18.2 this is illustrated as a "guide" moving along a thin springy rod and displacing the rod as it moves. Such a moving device generates a wave in the dispersive medium. The frequency of the wave is such that the phase velocity v of the wave matches the velocity v of the moving device. If the group velocity is less than the phase velocity, the wave that is generated trails behind the moving device, as shown in Figure 18.2. If the group velocity is greater than the phase velocity, the wave rushes out ahead of the moving device. Thus, an object that moves in a straight line at a constant velocity can radiate waves if the velocity of motion is equal to the phase velocity of the waves that are generated. This can occur in a linear dispersive medium, as we have noted

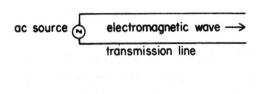

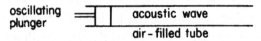

Figure 18.1

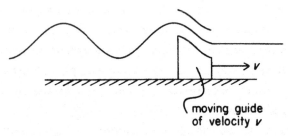

Figure 18.2

above. It can also occur in the case of an object moving through a space in which plane waves can travel.

Figure 18.3 depicts a medium in which plane waves travel with a velocity v. Imagine that the wavefront makes an angle θ with the z axis. Then the intersection of the wavefront on the z axis travels along the z axis with a velocity u given by

$$u = \frac{v}{\sin \theta}. \tag{18.1}$$

The easiest way of seeing this is that the component of u normal to the wavefront is $u \sin \theta$. We see that (18.1) is obviously true for $\theta = \pi/2$; in this case the wavefront is moving along normal to the z axis. Clearly also, as θ approaches zero, a slight motion

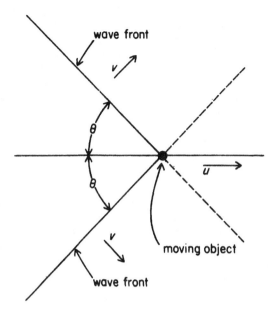

Figure 18.3

of the wavefront causes a large motion of its intersection with the z axis.

Let us now consider an object that moves through a medium that can support plane waves with a velocity v. When the velocity u of the object is greater than the velocity v of a plane wave, a plane wave traveling at an angle with the axis can just keep in step with the object. Thus, the object radiates; it excites waves whose wavefronts are inclined with respect to the axis at the angle θ given by (18.1).

When any solid object moves through air at a velocity greater than the velocity of sound, the object generates shock waves. Close to the object the behavior of the medium is nonlinear and the linear wave equation does not hold. But, farther from the object the waves travel out at the angle given by (18.1).

Objects cannot travel faster than the velocity of electromagnetic waves in vacuum (186,000 miles/second or 3×10^8 meters/second). Objects *can* travel faster than the velocity of electromagnetic waves in solids through which eletromagnetic waves can travel.

Many charged particles have a velocity greater than the velocity of light in glass, water, or transparent crystals. Cosmic rays, nuclear events, and man-made accelerators are sources of such particles. When such fast, charged particles travel through transparent solids they emit *Cherenkov radiation*. This accounts for the blue glow in water surrounding an atomic pile. The direction of Cherenkov radiation is specified by Equation 18.1.

When an object moves through a medium with a velocity greater than the velocity of plane waves in that medium, it continually gives up energy to the medium through the radiation of waves, and power must continually be supplied to keep the object in motion. Of course, power is needed for other reasons.

Physical media are lossy or *viscous*. It takes energy to drive a blimp through the air or a submarine through water even when the velocity of the blimp or the submarine is far below the velocity of sound in air or water. When an airplane flies, part of the power used to drive it is viscous or frictional loss as it slips through the air. Part of the power is spent in setting the air in motion in such a way as to support the airplane. However, the loss of energy through radiation of waves becomes important in supersonic aircraft and in boats.

Tiny ripples on the surface of water involve the mass of the water and the surface tension of the water. Such tiny ripples are nondispersive. Larger waves (in deep water) involve the mass of the water and its weight, which is mass times the acceleration of gravity. The dispersion relation for plane (really, linear) waves on the surface of the water is

$$\omega = (gk)^{1/2}. \tag{18.2}$$

The corresponding relation between phase velocity v and ω is

$$v = \frac{g}{\omega}. \tag{18.3}$$

Here g is the acceleration of gravity,

$g = 32$ feet/second$^2 = 9.6$ meters/second^2.

We see from (18.3) that a moving boat always has a velocity greater than a wave of *some* frequency. Hence, a boat moving on the surface of the water always generates waves, called a *wake*. The energy lost to such waves inceases rapidly with the speed of the boat.

Indeed, the only vehicles that can travel swiftly over the surface of the water are not deeply submerged in the water. They are either hydroplanes, which skitter over the surface, or hydrofoil boats, which are supported by small vanes under the surface, or air-cushion vehicles, which ride above the surface. Such vehicles produce smaller waves or wakes than vessels that float in the water.

Well streamlined submarines can travel swiftly through water. They are too far from the surface to excite a wake, and they travel at speeds far less than the velocity of sound in water.

In this book we have considered traveling disturbances which we call waves. Such waves travel away from the source that produces them.

There can also be unchanging disturbances in media that can support waves. A rod can support longitudinal sound waves; it can also be compressed by applying pressure at its ends, or it can be set in motion at a constant velocity. The air in a tube can support a sound wave; it can also be compressed (if the ends of the tube are closed), or there can be a steady flow of air through the tube. A stretched string can support transverse waves; it can also be forced into various shapes. A pair of wires can support an electromagnetic wave. But, we can maintain a constant, unvarying voltage between a pair of wires by connecting them to a battery. Or, if the wires are connected together at the far end, we can maintain a constant current in the wires.

A submarine that is motionless or that moves slower than the speed of sound does not send out waves. A bullet that moves faster than the speed of sound does send out waves.

We feel that there must be some relation between waves and disturbances which do not travel away from their source. Indeed, there is, and we can understand this relationship through examples.

Consider a wave sent along a stretched string that is fixed at the far end, as shown in Figure 18.4. The wave is reflected back toward the source. Along the string we have two waves of equal strength and the same frequency, one traveling to the right and the other traveling to the left.

In most instances we can use any appropriate physical

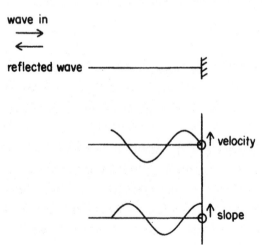

Figure 18.4

quantity as the measure of the strength S of a wave. However, as we have noted, all waves involve more than one sort of physical disturbance. Waves in stretched strings involve motion of the string as expressed by a small velocity which is alternately up and down as the wave travels. Such waves also involve a change of the slope of the string, which alternately increases and decreases as the wave travels. The velocity is greatest when the slope is greatest. *But*, for a positive slope, the velocity is upward for a wave traveling to the right and downward for a wave traveling to the left. What is the consequence of this when equal waves travel in opposite directions in the same medium?

Let us consider a wave traveling down a string with a fixed end, as shown in Figure 18.4. At the fixed end, the wave is reflected completely, so that the peak slope and velocity of the wave traveling to the left is equal to the peak slope and velocity of the wave traveling to the right. Clearly, at the fixed end the total velocity of the string must be zero. That is, the velocity associated with the wave moving to the right must be equal and opposite to the velocity associated with the wave moving to the left. If $z = 0$ at the fixed end, we can satisfy this if the total velocity, which we shall call V, is given by

$$V = \frac{V_0}{2j}\left(e^{j[\omega t - (\omega/v)z]} - e^{j[\omega t + (\omega/v)z]}\right)$$

$$= -V_0 e^{j\omega t} \sin[(\omega/v)z].$$

(18.4)

If the velocities associated with the waves traveling in opposite directions are equal and opposite at $z = 0$, the slopes are equal. Thus, we can express the total slope S as

$$S = \frac{S_0}{2}\left(e^{j[\omega t - (\omega/v)z]} + e^{j[\omega t + (\omega/v)z]}\right)$$

$$= S_0 e^{j\omega t} \cos[(\omega/v)z].$$

(18.5)

As illustrated by (18.4) and (18.5), two equal waves traveling in opposite directions form a standing wave. At any z position both the slope and the velocity vary sinusoidally with time. The peak amplitude of the slope and velocity vary sinusoidally with distance. Where the velocity V has the greatest peak amplitude the slope S is zero. Where the slope has the greatest peak amplitude the velocity V is zero.

This sort of relation is characteristic of all standing waves. In an electromagnetic standing wave, for example, the electric field is greatest where the magnetic field is zero, and the magnetic field is greatest where the electric field is zero.

We can think of the frequency of a standing wave as being as low as we wish. When the frequency goes to zero, we have a constant slope and velocity, or constant electric and magnetic fields which change neither with time nor with distance in the direction of propagation. We can regard these physical quantities as pertaining to two zero-frequency waves, one traveling to the left and the other traveling to the right. The ratio of the slope to the velocity, or of the electric field to the magnetic field, depends on the relative phases of the two waves. For certain phases the slope or the velocity can be zero; for certain phases the electric or the magnetic field can be zero. We can apply similar considerations to spherical waves and to waves of some other configurations.

What is the use of this representation of constant physical quantities such as pressure and velocity or electric and magnetic fields in terms of standing waves? In part, it provides an interesting conceptual link between static and dynamic problems. In part, however, in many cases it can be used to obtain results that apply to excitations of a medium which change with time but which are still associated with and "attached to" a moving source.

Consider the simple situation shown in Figure 18.5. To the left and right of two supports a stretched string has zero slope. Between the supports it has a slope $-S$. What happens when the two supports move to the right with a velocity u?

Now consider Figure 18.6. Here we have an object pushed by a force f and moving to the right with a velocity u. A wave of power P_1, energy per unit length E_1, and phase velocity v travels left toward the object. The moving object converts this incident wave into a wave of power P_2, energy per unit length E_2, and phase velocity v which travels from the moving object to the

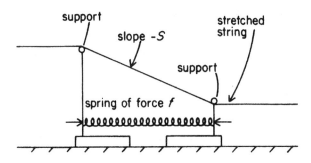

Figure 18.5

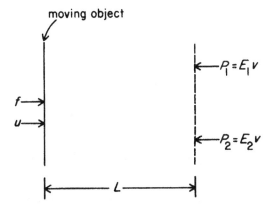

Figure 18.6

right. We assume nondispersive waves, so that v is both the phase and the group velocity of the waves. Under these circumstances the momentum per unit length p and the momentum flow b are

$$p = \frac{P}{v^2} = \frac{E}{v},$$

$$b = \frac{P}{v} = E. \tag{18.6}$$

The momentum and power of wave 1 are directed to the left; the momentum and power of wave 2 are directed to the right. Considering the waves to the right of a fixed reference point (the dashed vertical line) which is a changing distance L to the right of the moving object, there is a continual change of momentum which is the difference between the momentum of wave 1 and that of wave 2. This continous change of momentum is

$$\frac{1}{v}(P_1 + P_2) = E_1 + E_2. \tag{18.7}$$

Further, there is a certain amount of momentum stored in between the moving object and the fixed reference plane. This amount of momentum is

$$\frac{L}{v^2}(-P_1 + P_2) = \frac{L}{v}(-E_1 + E_2). \tag{18.8}$$

The rate of change of L is $-u$. Hence, the rate of change of the stored momentum is

$$\frac{u}{v}(E_1 - E_2). \tag{18.9}$$

The force f is responsible for the changes in momentum of (18.7) and (18.9). Hence

$$f = (E_1 + E_2) + \frac{u}{v}(E_1 - E_2). \tag{18.10}$$

The difference between the power P_2 and the power P_1 is

$$P_2 - P_1 = -v(E_1 - E_2). \tag{18.11}$$

The wave energy stored between the moving object and the fixed reference point is

$$L(E_1 + E_2). \tag{18.12}$$

The rate of change of this energy is

$$-u(E_1 + E_2). \tag{18.13}$$

The force f times the velocity u must be equal to the sum of the powers of (18.11) and (18.13), so that

$$fu = -u(E_1 + E_2) - v(E_1 - E_2),$$
$$f = -(E_1 + E_2) - \frac{v}{u}(E_1 - E_2). \tag{18.14}$$

We can easily obtain two relations from (18.10), which represents conservation of momentum, and (18.14), which represents conservation of energy. Let

$$E = E_1 + E_2.$$

Then

$$E = f\frac{[1 + (\frac{u}{v})^2]}{[1 - (\frac{u}{v})^2]}, \tag{18.15}$$

$$\frac{E_2}{E_1} = \frac{(1 + \frac{u}{v})^2}{(1 - \frac{u}{v})^2}. \tag{18.16}$$

Let us consider (18.15) and (18.16). Suppose we change the sign of u, so that the object moves away from the fixed reference point. The relation between force and total energy per unit length, (18.15), remains the same. The ratio between the energy of the wave traveling toward the moving object and the energy of the wave traveling away from the object is inverted. Thus,

(18.15) and (18.16) apply to the situation shown in Figure 18.7, in which a moving structure pushes on the waves between two objects with a force *f*, a force exerted to the right by the left-hand object, and a force exerted to the left by the right-hand object.

This is just the situation envisaged in Figure 18.5. Thus, we are in a position to explore what happens when a device that exerts a constant pressure on a section of a stretched string moves along the string.

When the device moves to the right, the energy of the wave traveling to the right becomes greater than the energy of the wave traveling to the left. Further, for a given force the total energy associated with the waves, E, becomes larger.

As the velocity u of the traveling device approaches the velocity v of the wave, the wave going to the right becomes much stronger than the wave going to the left. Further, the total energy of E becomes very large. This means that if we increase the velocity of the device of Figure 18.5, or Figure 18.7, we must exert a force on the device.

A part of the energy E can be interpreted as kinetic energy associated with the velocity u. If we let E_0 be the energy corresponding to zero velocity, then the energy associated with motion is

$$E - E_0 = \frac{2f(\frac{u}{v})^2}{1 - (\frac{u}{v})^2}. \tag{18.17}$$

At low velocities this kinetic energy is proportional to u^2, but the energy increases much more rapidly than u^2 as u approaches v.

We should note further that the increase in energy with velocity is associated with an increase in the magnitude of the slope, and hence as the velocity is increased the supports of Figure 18.5 move closer together. Some of the energy of (18.17)

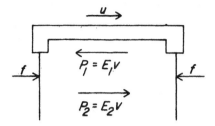

Figure 18.7

comes from the spring of the device of Figure 18.5 which shortens as the velocity of the device increases, but some must be supplied by pushing on the device in order to increase its velocity.

Here we have examined a very simple case of a disturbance that is carried along with a moving source.

For very low velocities, the disturbance is not very wave like. The stored energy is all of one sort, rather than being equally associated with slope and velocity, or with electric and magnetic fields.

As the velocity of the moving source is increased, the disturbance becomes more and more like a wave moving in the direction of motion of the source. For example, the fields that surround an electron that moves nearly as fast as light are very much like a very short intense electromagnetic wave. If the source could attain the wave velocity, it would produce a real forward-moving wave.

As the source approaches the wave velocity, the energy of the disturbance it produces becomes very large. If we took the equations literally, we would believe that it would take infinite energy to give the source the wave velocity. This is literally true for charged particles moving in a vacuum. Charged particles moving in solids and airplanes moving in the air can surpass the wave velocity with a finite expense of energy because the wave equation we use to describe electric and magnetic disturbances in a solid or acoustic disturbances in air does not hold for waves of very large strength.

We have also noted that the device of Figure 18.5 decreases in length as it increases in velocity. Even this has a counterpart in three-dimensional behavior. An electron and the electric and magnetic field patterns about it contract in the direction of motion as its velocity increases. This is a relativistic effect, but it was noted by Lorentz in pre-relativistic days. Material objects such as blimps or submarines do not contract in the direction of motion as their velocity increases, but the disturbances they produce in the medium do (ideally) contract in the direction of motion. We can see this from the scalar wave equation:

$$\frac{\partial^2 S}{\partial x^2} + \frac{\partial^2 S}{\partial y^2} + \frac{\partial^2 S}{\partial z^2} = \frac{1}{v^2}\frac{\partial^2 S}{\partial t^2}. \tag{18.18}$$

Let us assume that a solution of this equation is of the form

$$S = F(x, y, z, -ut).$$

That is, the disturbance in the medium moves to the right with a velocity u. For this sort of solution, (18.18) becomes

$$\frac{\partial^2 S}{\partial x^2} + \frac{\partial^2 S}{\partial y^2} + \left(1 - \frac{u^2}{v^2}\right)\frac{\partial^2 S}{\partial z^2} = 0. \tag{18.19}$$

Let

$$w = \frac{z}{\left(1 - \dfrac{u^2}{v^2}\right)^{1/2}}. \tag{18.20}$$

Then (18.19) becomes

$$\frac{\partial^2 S}{\partial x^2} + \frac{\partial^2 S}{\partial y^2} + \frac{\partial^2 S}{\partial w^2} = 0. \tag{18.21}$$

Equation (18.21) is of the same form as (18.18) would be for the case in which S does not change with time. Thus, for every *unmoving* type of disturbance in the medium there is a corresponding *moving* disturbance that satisfies w. For example, a time-independent solution of (18.18) is

$$S = \frac{1}{(x^2 + y^2 + z^2)^{1/2}}. \tag{18.22}$$

A solution of (18.21) is

$$\begin{aligned}
S &= \frac{1}{(x^2 + y^2 + w^2)^{1/2}} \\
&= \frac{1}{\{x^2 + y^2 + z^2/[1 - (u/v)^2]\}^{1/2}}.
\end{aligned} \tag{18.23}$$

The stationary disturbance of (18.22) is spherically symmetrical. The moving disturbance of (18.23) is flattened in the z direction, and becomes plane when $u = v$.

If the moving object contracts longitudinally as $[1 - (u/v)^2]^{1/2}$, it is easy to solve the equations for the disturbance it carries with it. Lorentz assumed that moving electrons so contracted with increasing velocity; Einstein later established this.[1] Lorentz's contemporary, Abraham, assumed that the moving electron remained spherical. It is much harder to treat this case, as it is hard to treat the case of moving blimps or submarines whose lengths do not change with velocity. The work of Lorentz and Abraham is discussed in Chapter XIX of Jean's *Electricity and Magnetism*.[2]

In reviewing what has been said so far, we may note that in the discussions associated with Figures 18.5–18.7 we have made calculations concerning *transverse* waves. Sound waves are longitudinal waves. A constant increase in pressure is a solution of the wave equation. In some sense we can think of a constant increase in pressure as a standing sound wave of zero frequency.

However, (18.15) does not give the correct relation between the force due to the pressure and the energy stored in the medium. There is a force at the boundary of a compressed medium that is proportional to pressure, rather than to energy, which is proportional to pressure squared.

We have noted that a source moving in a medium with a velocity less than the wave velocity does not radiate. It carries with it a disturbance in the medium. The time and space variation of the disturbance depend on the nature and velocity of the source.

When we change the velocity of the moving source, a new pattern eventually results, which has a different space and time variation and a different energy. At points remote from the moving source the pattern cannot change instantly when the velocity of the source changes. When the velocity of the source is changed, a part of the disturbances surrounding it is "shaken off" and travels as a radiated wave. This radiated wave moves past the moving source and acts on it. The force of the radiated field on the source is the force that we experience in changing the velocity of a moving source. If we change the velocity slowly, the energy radiated is negligible, but the radiated wave is still responsible for the force.[3]

We have discussed the phenomena associated with moving sources in terms of disturbances in the medium, disturbances that move along with the moving source. The nature of such disturbances is somehow intermediate between the nature of unvarying pressures, velocities, and fields and the nature of waves. Such disturbances can act on a source itself. Such action gives rise to an effective mass, as the effective mass of a body moving through water, or the *electromagnetic mass* of a charge moving through vacuum. Such self-action becomes very large as the speed of motion approaches the wave speed.

The disturbances carried along with moving bodies can also act on other bodies, as in the attraction between two bodies moving side by side through a fluid, or the magnetic attraction between electrons in parallel motion, an attraction that becomes just equal to the electrostatic repulsion between the electrons at the speed of light.

In quantum mechanics the action of disturbances that are carried along with sources are described in terms of *virtual photons* in the case of electromagnetic disturbances and in terms of *virtual phonons* in the case of vibrational disturbances. Thus, it seems appropriate to describe the disturbances that accompany a source which moves through a medium with a constant speed less than the wave velocity as a *virtual wave*.

References

1. C. Møller, *The Theory of Relativity*, New York: Oxford University Press, 1952.

2. Sir James H. Jeans, *Electricity and Magnetism*, Fifth Edition, New York: Cambridge University Press, 1933.

3. J. R. Pierce, Interaction of Moving Charges with Wave Circuits, *Journal of Applied Physics*, Vol. 26, pp. 627–638 (May, 1955).

Problems

1. Along the lines of Figure 18.4, describe reflection of a wave from a string that terminates in a massless vertical slider that is free to move vertically.

2. Along the lines of Figure 18.4, describe the reflection of an electromagnetic wave from a conductor.

3. A plasma with no thermal velocities oscillates with a particular frequency ω_p and waves in it have zero group velocity. What happens when a charged particle moves through such a plasma?

4. For a particular dispersive medium, the radian frequency ω and the wave vector k off a mode are related by

$$\omega = 1000k^2.$$

Units are meters and seconds. Suppose that an unfluctuating disturbance moves to the right along the medium with a velocity of 100 meters/second and sets up a wave. What will ω be? Will the wave be on the left or the right side of the moving disturbance?

5. Find a solution of (18.19) other than (18.24). Discuss it.

19 Retrospect and Prospect

There seems to be no end to things that can be known, but there is an end to what can be put into one book. My aim has been to discuss aspects of waves that are general and important and that can be explained with very little mathematics. Having learned this much about waves, some readers may wish to hear more of their practical uses, particularly in communications. Here, the *Encyclopedia Britannica*, and the *Americana*, and various easily digested books will be of help.[1-3]

Some readers may prefer to learn more about the detailed behavior of various sorts of waves; they may want to branch out from that which has been said to that which has been left unsaid.

In this book the reader has been introduced to waves as modes that travel in some medium, modes that vary sinusoidally with time and with distance in the direction of propagation. This approach is applicable to *all* unattenuated waves. Yet, the curious reader will want to see how such modes can be found through the solution of the partial differential equations for various physical systems, and how such modes are combined in fitting various initial or boundary conditions. Here I suggest encyclopedia articles and various useful books.[3-6]

If we review what we have covered, we find that the variety is wide and the applications are varied. In Chapter 1 we encountered the pitch of instruments and some aspects of the tempered scale. This opens up the whole world of sound and music—a world that can be explored through encyclopedia articles, books,[6-8] and the *Journal of the Acoustical Society of America*.

In Chapter 4 we came to the important concept of group velocity, the velocity with which the *envelope* of a wave train travels and also the speed with which the energy that constitutes the wave moves. We saw that group velocity varies with frequency, so that short pulses lengthen or disperse as they travel. Here we have a problem vital in the transmission of television and data signals.[9]

In Chapters 3 and 4 we learned about the behavior of coupled modes. This is important in practical devices such as the directional couplers used in microwave systems,[10] but it is even more important as a phenomenon which occurs in many contexts. We saw in Chapter 8 that when waves are coupled periodically, waves with different wave vectors or phase constants can interact strongly. If a transmission line or waveguide has periodic variations, it will pass waves of some frequencies and reflect waves of other frequencies. The periodic structure of a crystal lattice causes it to reflect electrons of some frequencies (or energies) and to allow electrons of other frequencies or energies to pass freely through it. This is important in the operation of transistors and other solid-state devices, and in much else besides. Those who wish may pursue such coupling or reflection in three dimensions rather than one, and learn more about the mature of the energy bands for electrons in crystalline materials of various sorts.[11]

Coupling between modes also offers a novel approach to the propagation of waves in tapered media such as horns, which is treated somewhat differently elsewhere.[5] As the amount of tapering is increased, some of the power put into a horn is reflected back. Thus, in such a tapered medium the power is not the total energy stored per unit length times the group velocity.

A tapered medium acts as a low-pass filter. When the frequency is lowered sufficiently (that is, when the wavelength is long enough), a wave does not travel in a horn; it is reflected. In an exponential horn the frequency at which reflection occurs is the same for all parts of the horn; in a conical horn, *cutoff*, or failure of a wave to travel through the horn, occurs at the small end. The radiation from the open end of a horn—or from the cone of a speaker—is also frequency sensitive, and that is why large apertures are necessary for efficient radiation of low frequencies.

In Chapter 7 we had noted that waves have momentum. In Chapter 10 we came to ascribe a momentum per unit length $p = E/v$ and a momentum flow $b = P/v$ to waves, where v is phase velocity, E is energy per unit length, and P is power. But, we found that this is not always the real, physical momentum. Treating p and b as momentum gives the right forces when a moving source or a wave acts to set up a wave, but part of the force may be exerted on the medium rather than transferred to the wave.

The treatment of energy and momentum lead to questions of energy and momentum of waves on a moving medium. Here we encountered the surprising fact that a wave can have, or can seem to have, a negative energy. That is, if we take energy away from the wave, its amplitude increases. This occurs when the medium moves fast enough so that a wave that moves to the left with respect to the medium moves to the right with respect to a stationary observer. This is just the case of the traveling wave tube, which is used to amplify microwave signals in communication satellites, and the reader may wish to learn more about this ingenious device .[12–14]

In Chapter 12, in studying energy and angular momentum in rotating media, we rather stumbled onto another useful device, the parametric amplifier. We explored other forms of parametric amplifiers in Chapter 13. In a parametric amplifier a strong wave pushes on weak waves and transfers energy and momentum to them. Parametric amplifiers are important in the frequency range of microwaves and also in the frequency range of light, and the reader may be interested in pursuing this further.[15]

In considering polarized waves we found how it is possible to convert between linearly and circularly polarized waves, and from one circular polarization to another. This led to a discussion of that strange and useful device, the isolator, which transmits waves traveling in one direction and absorbs waves traveling in the opposite direction. A variety of phenomena concerned with polarization are worth exploring.[10,16]

In Chapter 15 we came to consider plane and nearly plane waves and the radiation from complicated sources. Together with Chapter 16, this leads us to antenna theory. The reader may wish to compare the treatment of diffraction which is given in this book with the induction theorem of Schelkunoff,[5] which applies to electromagnetic waves, and he may wish to consider radio antennas in more detail, including the directive patterns of antennas.[17]

Chapter 17 brought us to the scalar wave equation—a point at which some books on waves start. Here the reader may wish to pursue the differences between scalar and vector waves.[4,5] He may wish to pursue a vector analog of Equation 16.22, which gives the largest intensity of a converging scalar wave intensity. It is clear that (16.22) can apply to pressure but not the velocity of a sound wave, and can apply to polarized waves for small angles only.

In Chapter 18 we encountered phenomena associated with sources in uniform motion—radiation when the velocity is greater than the wave velocity, and an added mass when the velocity is less than the wave velocity. This 'leads us in various directions; to a consideration of electromagnetic mass and special relativity on the one hand[18] and to problems of fluid dynamics on the other.[19]

I should note that this book has said almost nothing about

nonlinear waves, a subject that some readers may wish to pursue.[20]

Commonly, physicists and engineers first encounter waves in various complicated physical contexts and finally find the simple features that all waves have in common. Here the reader has considered those simple, common features, and is prepared, I hope, to see them exemplified in the complexities of the many areas I have pointed to, and in other areas as well. In so doing, the reader may consult advanced books. Or, he may find it profitable to push ahead on his own. The problems at the ends of the chapters can be of help. If some seem rather general or obscure, that is because they are intended not to recapitulate the text but to lead the reader beyond it.

Once started on his way, any ingenious person can make discoveries and inventions which are new to him, if not to man's science and technology. It's sometimes easier to discover something for oneself than it is to dig it out of a book, and when one can, it's certainly more enlightening and more fun.

References

1. A. H. Beck, *Words and Waves*, New York: World Universal Library, McGraw-Hill Book Company, 1967. (Simple story and explanations of applications in communication.)

2. J. R. Pierce, *Electrons and Waves*, Garden City, N. Y.: Anchor Books, Doubleday and Company, Inc., 1964. (Contains a simple account of Maxwell's equations.)

3. J. R. Pierce, *Waves and Messages*, Garden City, N. Y.: Anchor Books, Doubleday and Company, Inc., 1967. (Tells about telephony, television, communication satellites, and so on.)

4. William C. Elmore and Mark A. Heald, *Physics of Waves*,

New York: McGraw-Hill Book Company, 1969. (Contains a lot about mechanical and electromagnetic waves on an advanced undergraduate level.)*

5. Sergei A. Schelkunoff, *Electromagnetic Waves*, New York: D. Van Nostrand Company, Inc., 1943. (A large variety of methods for solving a large variety of problems.)

6. Philip M. Morse, *Vibration and Sound*, New York: McGraw-Hill Company, Inc., 1948. (An extensive text and reference book.)

7. John R. Pierce and Edward E. David, Jr., *Man's World of Sound*, Garden City, N. Y.: Doubleday and Company, Inc., 1958. (A simple book about sound, speech, and hearing.)

8. Herman L. F. Helmholtz, *On the Sensations of Tone*, New York: Dover Publications, Inc., 1954; translation of the Fourth German Edition of 1877. (A book by a great and musically talented scientist.)

9. Hendrik W. Bode, *Network Analysis and Feedback Amplifier Design*, New York: D. Van Nostrand Company, 1945. (Chapter XII discusses equalizers, a subject singularly absent in many texts about communication.)

10. Edward L. Ginzton, *Microwave Measurements*, New York: McGraw-Hill Book Company, Inc., 1957. (Contains a surprising amount of microwave lore.)

11. William Shockley, *Electrons and Holes in Semiconductors*, New York: D. Van Nostrand, Inc., 1950.

---

* Available from Dover Publications: ISBN 0-486-64926-1.
Visit www.doverpublications.com for availability and price.

12. R. Kompfner, *The Invention of the Traveling Wave Tube*, San Francisco, Calif.: San Francisco Press, Inc., 1964.

13. J. R. Pierce, *Traveling Wave Tubes*, New York: D. Van Nostrand Company, Inc., 1950.

14. A. W. H. Beck, *Space-Charge Waves*, New York: Pergamon Press, 1958.

15. Amnon Yariv, *Introduction to Optical Electronics*, New York: Holt, Rinehart and Winston, Inc., 1971.

16. William S. Shurcliff, *Polarized Light, Production and Use*, Cambridge, Mass.: Harvard University Press, 1962.

17. Sergei A. Schelkunoff and Harold T. Friis, *Antennas, Theory and Practice*, New York: John Wiley & Sons, Inc., 1952.

18. J. H. Jeans, *Electricity and Magnetism*, Fifth Edition, New York: Cambridge· University Press, 1933. (Chapters XIX and XX are close enough to early work on moving charges and on relativity to give the flavor of discovery.)

19. Sir Horace Lamb, *Hydrodynamics*, New York: Dover Publications, 1945. (Tells about effective mass of objects moving through a fluid, but for small velocities only.)

20. Alan Jeffery, *Nonlinear Wave Propagation, with Applications to Physics and Magnetohydrodynamics*, New York: Academic Press, 1964.

Index

A CATALOG OF SELECTED
DOVER BOOKS
IN SCIENCE AND MATHEMATICS

Astronomy

CHARIOTS FOR APOLLO: The NASA History of Manned Lunar Spacecraft to 1969, Courtney G. Brooks, James M. Grimwood, and Loyd S. Swenson, Jr. This illustrated history by a trio of experts is the definitive reference on the Apollo spacecraft and lunar modules. It traces the vehicles' design, development, and operation in space. More than 100 photographs and illustrations. 576pp. 6 3/4 x 9 1/4. 0-486-46756-2

EXPLORING THE MOON THROUGH BINOCULARS AND SMALL TELESCOPES, Ernest H. Cherrington, Jr. Informative, profusely illustrated guide to locating and identifying craters, rills, seas, mountains, other lunar features. Newly revised and updated with special section of new photos. Over 100 photos and diagrams. 240pp. 8 1/4 x 11. 0-486-24491-1

WHERE NO MAN HAS GONE BEFORE: A History of NASA's Apollo Lunar Expeditions, William David Compton. Introduction by Paul Dickson. This official NASA history traces behind-the-scenes conflicts and cooperation between scientists and engineers. The first half concerns preparations for the Moon landings, and the second half documents the flights that followed Apollo 11. 1989 edition. 432pp. 7 x 10. 0-486-47888-2

APOLLO EXPEDITIONS TO THE MOON: The NASA History, Edited by Edgar M. Cortright. Official NASA publication marks the 40th anniversary of the first lunar landing and features essays by project participants recalling engineering and administrative challenges. Accessible, jargon-free accounts, highlighted by numerous illustrations. 336pp. 8 3/8 x 10 7/8. 0-486-47175-6

ON MARS: Exploration of the Red Planet, 1958-1978--The NASA History, Edward Clinton Ezell and Linda Neuman Ezell. NASA's official history chronicles the start of our explorations of our planetary neighbor. It recounts cooperation among government, industry, and academia, and it features dozens of photos from Viking cameras. 560pp. 6 3/4 x 9 1/4. 0-486-46757-0

ARISTARCHUS OF SAMOS: The Ancient Copernicus, Sir Thomas Heath. Heath's history of astronomy ranges from Homer and Hesiod to Aristarchus and includes quotes from numerous thinkers, compilers, and scholasticists from Thales and Anaximander through Pythagoras, Plato, Aristotle, and Heraclides. 34 figures. 448pp. 5 3/8 x 8 1/2. 0-486-43886-4

AN INTRODUCTION TO CELESTIAL MECHANICS, Forest Ray Moulton. Classic text still unsurpassed in presentation of fundamental principles. Covers rectilinear motion, central forces, problems of two and three bodies, much more. Includes over 200 problems, some with answers. 437pp. 5 3/8 x 8 1/2. 0-486-64687-4

BEYOND THE ATMOSPHERE: Early Years of Space Science, Homer E. Newell. This exciting survey is the work of a top NASA administrator who chronicles technological advances, the relationship of space science to general science, and the space program's social, political, and economic contexts. 528pp. 6 3/4 x 9 1/4. 0-486-47464-X

STAR LORE: Myths, Legends, and Facts, William Tyler Olcott. Captivating retellings of the origins and histories of ancient star groups include Pegasus, Ursa Major, Pleiades, signs of the zodiac, and other constellations. "Classic." – *Sky & Telescope.* 58 illustrations. 544pp. 5 3/8 x 8 1/2. 0-486-43581-4

A COMPLETE MANUAL OF AMATEUR ASTRONOMY: Tools and Techniques for Astronomical Observations, P. Clay Sherrod with Thomas L. Koed. Concise, highly readable book discusses the selection, set-up, and maintenance of a telescope; amateur studies of the sun; lunar topography and occultations; and more. 124 figures. 26 halftones. 37 tables. 335pp. 6 1/2 x 9 1/4. 0-486-42820-6

Browse over 9,000 books at www.doverpublications.com

Chemistry

MOLECULAR COLLISION THEORY, M. S. Child. This high-level monograph offers an analytical treatment of classical scattering by a central force, quantum scattering by a central force, elastic scattering phase shifts, and semi-classical elastic scattering. 1974 edition. 310pp. 5 3/8 x 8 1/2. 0-486-69437-2

HANDBOOK OF COMPUTATIONAL QUANTUM CHEMISTRY, David B. Cook. This comprehensive text provides upper-level undergraduates and graduate students with an accessible introduction to the implementation of quantum ideas in molecular modeling, exploring practical applications alongside theoretical explanations. 1998 edition. 832pp. 5 3/8 x 8 1/2. 0-486-44307-8

RADIOACTIVE SUBSTANCES, Marie Curie. The celebrated scientist's thesis, which directly preceded her 1903 Nobel Prize, discusses establishing atomic character of radioactivity; extraction from pitchblende of polonium and radium; isolation of pure radium chloride; more. 96pp. 5 3/8 x 8 1/2. 0-486-42550-9

CHEMICAL MAGIC, Leonard A. Ford. Classic guide provides intriguing entertainment while elucidating sound scientific principles, with more than 100 unusual stunts: cold fire, dust explosions, a nylon rope trick, a disappearing beaker, much more. 128pp. 5 3/8 x 8 1/2. 0-486-67628-5

ALCHEMY, E. J. Holmyard. Classic study by noted authority covers 2,000 years of alchemical history: religious, mystical overtones; apparatus; signs, symbols, and secret terms; advent of scientific method, much more. Illustrated. 320pp. 5 3/8 x 8 1/2.
0-486-26298-7

CHEMICAL KINETICS AND REACTION DYNAMICS, Paul L. Houston. This text teaches the principles underlying modern chemical kinetics in a clear, direct fashion, using several examples to enhance basic understanding. Solutions to selected problems. 2001 edition. 352pp. 8 3/8 x 11. 0-486-45334-0

PROBLEMS AND SOLUTIONS IN QUANTUM CHEMISTRY AND PHYSICS, Charles S. Johnson and Lee G. Pedersen. Unusually varied problems, with detailed solutions, cover of quantum mechanics, wave mechanics, angular momentum, molecular spectroscopy, scattering theory, more. 280 problems, plus 139 supplementary exercises. 430pp. 6 1/2 x 9 1/4. 0-486-65236-X

ELEMENTS OF CHEMISTRY, Antoine Lavoisier. Monumental classic by the founder of modern chemistry features first explicit statement of law of conservation of matter in chemical change, and more. Facsimile reprint of original (1790) Kerr translation. 539pp. 5 3/8 x 8 1/2. 0-486-64624-6

MAGNETISM AND TRANSITION METAL COMPLEXES, F. E. Mabbs and D. J. Machin. A detailed view of the calculation methods involved in the magnetic properties of transition metal complexes, this volume offers sufficient background for original work in the field. 1973 edition. 240pp. 5 3/8 x 8 1/2. 0-486-46284-6

GENERAL CHEMISTRY, Linus Pauling. Revised third edition of classic first-year text by Nobel laureate. Atomic and molecular structure, quantum mechanics, statistical mechanics, thermodynamics correlated with descriptive chemistry. Problems. 992pp. 5 3/8 x 8 1/2. 0-486-65622-5

ELECTROLYTE SOLUTIONS: Second Revised Edition, R. A. Robinson and R. H. Stokes. Classic text deals primarily with measurement, interpretation of conductance, chemical potential, and diffusion in electrolyte solutions. Detailed theoretical interpretations, plus extensive tables of thermodynamic and transport properties. 1970 edition. 590pp. 5 3/8 x 8 1/2. 0-486-42225-9

Browse over 9,000 books at www.doverpublications.com

Engineering

FUNDAMENTALS OF ASTRODYNAMICS, Roger R. Bate, Donald D. Mueller, and Jerry E. White. Teaching text developed by U.S. Air Force Academy develops the basic two-body and n-body equations of motion; orbit determination; classical orbital elements, coordinate transformations; differential correction; more. 1971 edition. 455pp. 5 3/8 x 8 1/2. 0-486-60061-0

INTRODUCTION TO CONTINUUM MECHANICS FOR ENGINEERS: Revised Edition, Ray M. Bowen. This self-contained text introduces classical continuum models within a modern framework. Its numerous exercises illustrate the governing principles, linearizations, and other approximations that constitute classical continuum models. 2007 edition. 320pp. 6 1/8 x 9 1/4. 0-486-47460-7

ENGINEERING MECHANICS FOR STRUCTURES, Louis L. Bucciarelli. This text explores the mechanics of solids and statics as well as the strength of materials and elasticity theory. Its many design exercises encourage creative initiative and systems thinking. 2009 edition. 320pp. 6 1/8 x 9 1/4. 0-486-46855-0

FEEDBACK CONTROL THEORY, John C. Doyle, Bruce A. Francis and Allen R. Tannenbaum. This excellent introduction to feedback control system design offers a theoretical approach that captures the essential issues and can be applied to a wide range of practical problems. 1992 edition. 224pp. 6 1/2 x 9 1/4. 0-486-46933-6

THE FORCES OF MATTER, Michael Faraday. These lectures by a famous inventor offer an easy-to-understand introduction to the interactions of the universe's physical forces. Six essays explore gravitation, cohesion, chemical affinity, heat, magnetism, and electricity. 1993 edition. 96pp. 5 3/8 x 8 1/2. 0-486-47482-8

DYNAMICS, Lawrence E. Goodman and William H. Warner. Beginning engineering text introduces calculus of vectors, particle motion, dynamics of particle systems and plane rigid bodies, technical applications in plane motions, and more. Exercises and answers in every chapter. 619pp. 5 3/8 x 8 1/2. 0-486-42006-X

ADAPTIVE FILTERING PREDICTION AND CONTROL, Graham C. Goodwin and Kwai Sang Sin. This unified survey focuses on linear discrete-time systems and explores natural extensions to nonlinear systems. It emphasizes discrete-time systems, summarizing theoretical and practical aspects of a large class of adaptive algorithms. 1984 edition. 560pp. 6 1/2 x 9 1/4. 0-486-46932-8

INDUCTANCE CALCULATIONS, Frederick W. Grover. This authoritative reference enables the design of virtually every type of inductor. It features a single simple formula for each type of inductor, together with tables containing essential numerical factors. 1946 edition. 304pp. 5 3/8 x 8 1/2. 0-486-47440-2

THERMODYNAMICS: Foundations and Applications, Elias P. Gyftopoulos and Gian Paolo Beretta. Designed by two MIT professors, this authoritative text discusses basic concepts and applications in detail, emphasizing generality, definitions, and logical consistency. More than 300 solved problems cover realistic energy systems and processes. 800pp. 6 1/8 x 9 1/4. 0-486-43932-1

THE FINITE ELEMENT METHOD: Linear Static and Dynamic Finite Element Analysis, Thomas J. R. Hughes. Text for students without in-depth mathematical training, this text includes a comprehensive presentation and analysis of algorithms of time-dependent phenomena plus beam, plate, and shell theories. Solution guide available upon request. 672pp. 6 1/2 x 9 1/4. 0-486-41181-8

Browse over 9,000 books at www.doverpublications.com

HELICOPTER THEORY, Wayne Johnson. Monumental engineering text covers vertical flight, forward flight, performance, mathematics of rotating systems, rotary wing dynamics and aerodynamics, aeroelasticity, stability and control, stall, noise, and more. 189 illustrations. 1980 edition. 1089pp. 5 5/8 x 8 1/4. 0-486-68230-7

MATHEMATICAL HANDBOOK FOR SCIENTISTS AND ENGINEERS: Definitions, Theorems, and Formulas for Reference and Review, Granino A. Korn and Theresa M. Korn. Convenient access to information from every area of mathematics: Fourier transforms, Z transforms, linear and nonlinear programming, calculus of variations, random-process theory, special functions, combinatorial analysis, game theory, much more. 1152pp. 5 3/8 x 8 1/2. 0-486-41147-8

A HEAT TRANSFER TEXTBOOK: Fourth Edition, John H. Lienhard V and John H. Lienhard IV. This introduction to heat and mass transfer for engineering students features worked examples and end-of-chapter exercises. Worked examples and end-of-chapter exercises appear throughout the book, along with well-drawn, illuminating figures. 768pp. 7 x 9 1/4. 0-486-47931-5

BASIC ELECTRICITY, U.S. Bureau of Naval Personnel. Originally a training course; best nontechnical coverage. Topics include batteries, circuits, conductors, AC and DC, inductance and capacitance, generators, motors, transformers, amplifiers, etc. Many questions with answers. 349 illustrations. 1969 edition. 448pp. 6 1/2 x 9 1/4.
0-486-20973-3

BASIC ELECTRONICS, U.S. Bureau of Naval Personnel. Clear, well-illustrated introduction to electronic equipment covers numerous essential topics: electron tubes, semiconductors, electronic power supplies, tuned circuits, amplifiers, receivers, ranging and navigation systems, computers, antennas, more. 560 illustrations. 567pp. 6 1/2 x 9 1/4. 0-486-21076-6

BASIC WING AND AIRFOIL THEORY, Alan Pope. This self-contained treatment by a pioneer in the study of wind effects covers flow functions, airfoil construction and pressure distribution, finite and monoplane wings, and many other subjects. 1951 edition. 320pp. 5 3/8 x 8 1/2. 0-486-47188-8

SYNTHETIC FUELS, Ronald F. Probstein and R. Edwin Hicks. This unified presentation examines the methods and processes for converting coal, oil, shale, tar sands, and various forms of biomass into liquid, gaseous, and clean solid fuels. 1982 edition. 512pp. 6 1/8 x 9 1/4. 0-486-44977-7

THEORY OF ELASTIC STABILITY, Stephen P. Timoshenko and James M. Gere. Written by world-renowned authorities on mechanics, this classic ranges from theoretical explanations of 2- and 3-D stress and strain to practical applications such as torsion, bending, and thermal stress. 1961 edition. 560pp. 5 3/8 x 8 1/2. 0-486-47207-8

PRINCIPLES OF DIGITAL COMMUNICATION AND CODING, Andrew J. Viterbi and Jim K. Omura. This classic by two digital communications experts is geared toward students of communications theory and to designers of channels, links, terminals, modems, or networks used to transmit and receive digital messages. 1979 edition. 576pp. 6 1/8 x 9 1/4. 0-486-46901-8

LINEAR SYSTEM THEORY: The State Space Approach, Lotfi A. Zadeh and Charles A. Desoer. Written by two pioneers in the field, this exploration of the state space approach focuses on problems of stability and control, plus connections between this approach and classical techniques. 1963 edition. 656pp. 6 1/8 x 9 1/4.
0-486-46663-9

Browse over 9,000 books at www.doverpublications.com

Mathematics–Bestsellers

HANDBOOK OF MATHEMATICAL FUNCTIONS: with Formulas, Graphs, and Mathematical Tables, Edited by Milton Abramowitz and Irene A. Stegun. A classic resource for working with special functions, standard trig, and exponential logarithmic definitions and extensions, it features 29 sets of tables, some to as high as 20 places. 1046pp. 8 x 10 1/2. 0-486-61272-4

ABSTRACT AND CONCRETE CATEGORIES: The Joy of Cats, Jiri Adamek, Horst Herrlich, and George E. Strecker. This up-to-date introductory treatment employs category theory to explore the theory of structures. Its unique approach stresses concrete categories and presents a systematic view of factorization structures. Numerous examples. 1990 edition, updated 2004. 528pp. 6 1/8 x 9 1/4. 0-486-46934-4

MATHEMATICS: Its Content, Methods and Meaning, A. D. Aleksandrov, A. N. Kolmogorov, and M. A. Lavrent'ev. Major survey offers comprehensive, coherent discussions of analytic geometry, algebra, differential equations, calculus of variations, functions of a complex variable, prime numbers, linear and non-Euclidean geometry, topology, functional analysis, more. 1963 edition. 1120pp. 5 3/8 x 8 1/2. 0-486-40916-3

INTRODUCTION TO VECTORS AND TENSORS: Second Edition--Two Volumes Bound as One, Ray M. Bowen and C.-C. Wang. Convenient single-volume compilation of two texts offers both introduction and in-depth survey. Geared toward engineering and science students rather than mathematicians, it focuses on physics and engineering applications. 1976 edition. 560pp. 6 1/2 x 9 1/4. 0-486-46914-X

AN INTRODUCTION TO ORTHOGONAL POLYNOMIALS, Theodore S. Chihara. Concise introduction covers general elementary theory, including the representation theorem and distribution functions, continued fractions and chain sequences, the recurrence formula, special functions, and some specific systems. 1978 edition. 272pp. 5 3/8 x 8 1/2. 0-486-47929-3

ADVANCED MATHEMATICS FOR ENGINEERS AND SCIENTISTS, Paul DuChateau. This primary text and supplemental reference focuses on linear algebra, calculus, and ordinary differential equations. Additional topics include partial differential equations and approximation methods. Includes solved problems. 1992 edition. 400pp. 7 1/2 x 9 1/4. 0-486-47930-7

PARTIAL DIFFERENTIAL EQUATIONS FOR SCIENTISTS AND ENGINEERS, Stanley J. Farlow. Practical text shows how to formulate and solve partial differential equations. Coverage of diffusion-type problems, hyperbolic-type problems, elliptic-type problems, numerical and approximate methods. Solution guide available upon request. 1982 edition. 414pp. 6 1/8 x 9 1/4. 0-486-67620-X

VARIATIONAL PRINCIPLES AND FREE-BOUNDARY PROBLEMS, Avner Friedman. Advanced graduate-level text examines variational methods in partial differential equations and illustrates their applications to free-boundary problems. Features detailed statements of standard theory of elliptic and parabolic operators. 1982 edition. 720pp. 6 1/8 x 9 1/4. 0-486-47853-X

LINEAR ANALYSIS AND REPRESENTATION THEORY, Steven A. Gaal. Unified treatment covers topics from the theory of operators and operator algebras on Hilbert spaces; integration and representation theory for topological groups; and the theory of Lie algebras, Lie groups, and transform groups. 1973 edition. 704pp. 6 1/8 x 9 1/4. 0-486-47851-3

Browse over 9,000 books at www.doverpublications.com

A SURVEY OF INDUSTRIAL MATHEMATICS, Charles R. MacCluer. Students learn how to solve problems they'll encounter in their professional lives with this concise single-volume treatment. It employs MATLAB and other strategies to explore typical industrial problems. 2000 edition. 384pp. 5 3/8 x 8 1/2.   0-486-47702-9

NUMBER SYSTEMS AND THE FOUNDATIONS OF ANALYSIS, Elliott Mendelson. Geared toward undergraduate and beginning graduate students, this study explores natural numbers, integers, rational numbers, real numbers, and complex numbers. Numerous exercises and appendixes supplement the text. 1973 edition. 368pp. 5 3/8 x 8 1/2.   0-486-45792-3

A FIRST LOOK AT NUMERICAL FUNCTIONAL ANALYSIS, W. W. Sawyer. Text by renowned educator shows how problems in numerical analysis lead to concepts of functional analysis. Topics include Banach and Hilbert spaces, contraction mappings, convergence, differentiation and integration, and Euclidean space. 1978 edition. 208pp. 5 3/8 x 8 1/2.   0-486-47882-3

FRACTALS, CHAOS, POWER LAWS: Minutes from an Infinite Paradise, Manfred Schroeder. A fascinating exploration of the connections between chaos theory, physics, biology, and mathematics, this book abounds in award-winning computer graphics, optical illusions, and games that clarify memorable insights into self-similarity. 1992 edition. 448pp. 6 1/8 x 9 1/4.   0-486-47204-3

SET THEORY AND THE CONTINUUM PROBLEM, Raymond M. Smullyan and Melvin Fitting. A lucid, elegant, and complete survey of set theory, this three-part treatment explores axiomatic set theory, the consistency of the continuum hypothesis, and forcing and independence results. 1996 edition. 336pp. 6 x 9.   0-486-47484-4

DYNAMICAL SYSTEMS, Shlomo Sternberg. A pioneer in the field of dynamical systems discusses one-dimensional dynamics, differential equations, random walks, iterated function systems, symbolic dynamics, and Markov chains. Supplementary materials include PowerPoint slides and MATLAB exercises. 2010 edition. 272pp. 6 1/8 x 9 1/4.   0-486-47705-3

ORDINARY DIFFERENTIAL EQUATIONS, Morris Tenenbaum and Harry Pollard. Skillfully organized introductory text examines origin of differential equations, then defines basic terms and outlines general solution of a differential equation. Explores integrating factors; dilution and accretion problems; Laplace Transforms; Newton's Interpolation Formulas, more. 818pp. 5 3/8 x 8 1/2.   0-486-64940-7

MATROID THEORY, D. J. A. Welsh. Text by a noted expert describes standard examples and investigation results, using elementary proofs to develop basic matroid properties before advancing to a more sophisticated treatment. Includes numerous exercises. 1976 edition. 448pp. 5 3/8 x 8 1/2.   0-486-47439-9

THE CONCEPT OF A RIEMANN SURFACE, Hermann Weyl. This classic on the general history of functions combines function theory and geometry, forming the basis of the modern approach to analysis, geometry, and topology. 1955 edition. 208pp. 5 3/8 x 8 1/2.   0-486-47004-0

THE LAPLACE TRANSFORM, David Vernon Widder. This volume focuses on the Laplace and Stieltjes transforms, offering a highly theoretical treatment. Topics include fundamental formulas, the moment problem, monotonic functions, and Tauberian theorems. 1941 edition. 416pp. 5 3/8 x 8 1/2.   0-486-47755-X

Mathematics–Logic and Problem Solving

PERPLEXING PUZZLES AND TANTALIZING TEASERS, Martin Gardner. Ninety-three riddles, mazes, illusions, tricky questions, word and picture puzzles, and other challenges offer hours of entertainment for youngsters. Filled with rib-tickling drawings. Solutions. 224pp. 5 3/8 x 8 1/2. 0-486-25637-5

MY BEST MATHEMATICAL AND LOGIC PUZZLES, Martin Gardner. The noted expert selects 70 of his favorite "short" puzzles. Includes The Returning Explorer, The Mutilated Chessboard, Scrambled Box Tops, and dozens more. Complete solutions included. 96pp. 5 3/8 x 8 1/2. 0-486-28152-3

THE LADY OR THE TIGER?: and Other Logic Puzzles, Raymond M. Smullyan. Created by a renowned puzzle master, these whimsically themed challenges involve paradoxes about probability, time, and change; metapuzzles; and self-referentiality. Nineteen chapters advance in difficulty from relatively simple to highly complex. 1982 edition. 240pp. 5 3/8 x 8 1/2. 0-486-47027-X

SATAN, CANTOR AND INFINITY: Mind-Boggling Puzzles, Raymond M. Smullyan. A renowned mathematician tells stories of knights and knaves in an entertaining look at the logical precepts behind infinity, probability, time, and change. Requires a strong background in mathematics. Complete solutions. 288pp. 5 3/8 x 8 1/2.

0-486-47036-9

THE RED BOOK OF MATHEMATICAL PROBLEMS, Kenneth S. Williams and Kenneth Hardy. Handy compilation of 100 practice problems, hints and solutions indispensable for students preparing for the William Lowell Putnam and other mathematical competitions. Preface to the First Edition. Sources. 1988 edition. 192pp. 5 3/8 x 8 1/2. 0-486-69415-1

KING ARTHUR IN SEARCH OF HIS DOG AND OTHER CURIOUS PUZZLES, Raymond M. Smullyan. This fanciful, original collection for readers of all ages features arithmetic puzzles, logic problems related to crime detection, and logic and arithmetic puzzles involving King Arthur and his Dogs of the Round Table. 160pp. 5 3/8 x 8 1/2.

0-486-47435-6

UNDECIDABLE THEORIES: Studies in Logic and the Foundation of Mathematics, Alfred Tarski in collaboration with Andrzej Mostowski and Raphael M. Robinson. This well-known book by the famed logician consists of three treatises: "A General Method in Proofs of Undecidability," "Undecidability and Essential Undecidability in Mathematics," and "Undecidability of the Elementary Theory of Groups." 1953 edition. 112pp. 5 3/8 x 8 1/2. 0-486-47703-7

LOGIC FOR MATHEMATICIANS, J. Barkley Rosser. Examination of essential topics and theorems assumes no background in logic. "Undoubtedly a major addition to the literature of mathematical logic." – *Bulletin of the American Mathematical Society.* 1978 edition. 592pp. 6 1/8 x 9 1/4. 0-486-46898-4

INTRODUCTION TO PROOF IN ABSTRACT MATHEMATICS, Andrew Wohlgemuth. This undergraduate text teaches students what constitutes an acceptable proof, and it develops their ability to do proofs of routine problems as well as those requiring creative insights. 1990 edition. 384pp. 6 1/2 x 9 1/4. 0-486-47854-8

FIRST COURSE IN MATHEMATICAL LOGIC, Patrick Suppes and Shirley Hill. Rigorous introduction is simple enough in presentation and context for wide range of students. Symbolizing sentences; logical inference; truth and validity; truth tables; terms, predicates, universal quantifiers; universal specification and laws of identity; more. 288pp. 5 3/8 x 8 1/2. 0-486-42259-3

Browse over 9,000 books at www.doverpublications.com

Mathematics–Algebra and Calculus

VECTOR CALCULUS, Peter Baxandall and Hans Liebeck. This introductory text offers a rigorous, comprehensive treatment. Classical theorems of vector calculus are amply illustrated with figures, worked examples, physical applications, and exercises with hints and answers. 1986 edition. 560pp. 5 3/8 x 8 1/2.   0-486-46620-5

ADVANCED CALCULUS: An Introduction to Classical Analysis, Louis Brand. A course in analysis that focuses on the functions of a real variable, this text introduces the basic concepts in their simplest setting and illustrates its teachings with numerous examples, theorems, and proofs. 1955 edition. 592pp. 5 3/8 x 8 1/2.   0-486-44548-8

ADVANCED CALCULUS, Avner Friedman. Intended for students who have already completed a one-year course in elementary calculus, this two-part treatment advances from functions of one variable to those of several variables. Solutions. 1971 edition. 432pp. 5 3/8 x 8 1/2.   0-486-45795-8

METHODS OF MATHEMATICS APPLIED TO CALCULUS, PROBABILITY, AND STATISTICS, Richard W. Hamming. This 4-part treatment begins with algebra and analytic geometry and proceeds to an exploration of the calculus of algebraic functions and transcendental functions and applications. 1985 edition. Includes 310 figures and 18 tables. 880pp. 6 1/2 x 9 1/4.   0-486-43945-3

BASIC ALGEBRA I: Second Edition, Nathan Jacobson. A classic text and standard reference for a generation, this volume covers all undergraduate algebra topics, including groups, rings, modules, Galois theory, polynomials, linear algebra, and associative algebra. 1985 edition. 528pp. 6 1/8 x 9 1/4.   0-486-47189-6

BASIC ALGEBRA II: Second Edition, Nathan Jacobson. This classic text and standard reference comprises all subjects of a first-year graduate-level course, including in-depth coverage of groups and polynomials and extensive use of categories and functors. 1989 edition. 704pp. 6 1/8 x 9 1/4.   0-486-47187-X

CALCULUS: An Intuitive and Physical Approach (Second Edition), Morris Kline. Application-oriented introduction relates the subject as closely as possible to science with explorations of the derivative; differentiation and integration of the powers of x; theorems on differentiation, antidifferentiation; the chain rule; trigonometric functions; more. Examples. 1967 edition. 960pp. 6 1/2 x 9 1/4.   0-486-40453-6

ABSTRACT ALGEBRA AND SOLUTION BY RADICALS, John E. Maxfield and Margaret W. Maxfield. Accessible advanced undergraduate-level text starts with groups, rings, fields, and polynomials and advances to Galois theory, radicals and roots of unity, and solution by radicals. Numerous examples, illustrations, exercises, appendixes. 1971 edition. 224pp. 6 1/8 x 9 1/4.   0-486-47723-1

AN INTRODUCTION TO THE THEORY OF LINEAR SPACES, Georgi E. Shilov. Translated by Richard A. Silverman. Introductory treatment offers a clear exposition of algebra, geometry, and analysis as parts of an integrated whole rather than separate subjects. Numerous examples illustrate many different fields, and problems include hints or answers. 1961 edition. 320pp. 5 3/8 x 8 1/2.   0-486-63070-6

LINEAR ALGEBRA, Georgi E. Shilov. Covers determinants, linear spaces, systems of linear equations, linear functions of a vector argument, coordinate transformations, the canonical form of the matrix of a linear operator, bilinear and quadratic forms, and more. 387pp. 5 3/8 x 8 1/2.   0-486-63518-X

Browse over 9,000 books at www.doverpublications.com

Mathematics–Probability and Statistics

BASIC PROBABILITY THEORY, Robert B. Ash. This text emphasizes the probabilistic way of thinking, rather than measure-theoretic concepts. Geared toward advanced undergraduates and graduate students, it features solutions to some of the problems. 1970 edition. 352pp. 5 3/8 x 8 1/2.
0-486-46628-0

PRINCIPLES OF STATISTICS, M. G. Bulmer. Concise description of classical statistics, from basic dice probabilities to modern regression analysis. Equal stress on theory and applications. Moderate difficulty; only basic calculus required. Includes problems with answers. 252pp. 5 5/8 x 8 1/4.
0-486-63760-3

OUTLINE OF BASIC STATISTICS: Dictionary and Formulas, John E. Freund and Frank J. Williams. Handy guide includes a 70-page outline of essential statistical formulas covering grouped and ungrouped data, finite populations, probability, and more, plus over 1,000 clear, concise definitions of statistical terms. 1966 edition. 208pp. 5 3/8 x 8 1/2.
0-486-47769-X

GOOD THINKING: The Foundations of Probability and Its Applications, Irving J. Good. This in-depth treatment of probability theory by a famous British statistician explores Keynesian principles and surveys such topics as Bayesian rationality, corroboration, hypothesis testing, and mathematical tools for induction and simplicity. 1983 edition. 352pp. 5 3/8 x 8 1/2.
0-486-47438-0

INTRODUCTION TO PROBABILITY THEORY WITH CONTEMPORARY APPLICATIONS, Lester L. Helms. Extensive discussions and clear examples, written in plain language, expose students to the rules and methods of probability. Exercises foster problem-solving skills, and all problems feature step-by-step solutions. 1997 edition. 368pp. 6 1/2 x 9 1/4.
0-486-47418-6

CHANCE, LUCK, AND STATISTICS, Horace C. Levinson. In simple, non-technical language, this volume explores the fundamentals governing chance and applies them to sports, government, and business. "Clear and lively ... remarkably accurate." – *Scientific Monthly*. 384pp. 5 3/8 x 8 1/2.
0-486-41997-5

FIFTY CHALLENGING PROBLEMS IN PROBABILITY WITH SOLUTIONS, Frederick Mosteller. Remarkable puzzlers, graded in difficulty, illustrate elementary and advanced aspects of probability. These problems were selected for originality, general interest, or because they demonstrate valuable techniques. Also includes detailed solutions. 88pp. 5 3/8 x 8 1/2.
0-486-65355-2

EXPERIMENTAL STATISTICS, Mary Gibbons Natrella. A handbook for those seeking engineering information and quantitative data for designing, developing, constructing, and testing equipment. Covers the planning of experiments, the analyzing of extreme-value data; and more. 1966 edition. Index. Includes 52 figures and 76 tables. 560pp. 8 3/8 x 11.
0-486-43937-2

STOCHASTIC MODELING: Analysis and Simulation, Barry L. Nelson. Coherent introduction to techniques also offers a guide to the mathematical, numerical, and simulation tools of systems analysis. Includes formulation of models, analysis, and interpretation of results. 1995 edition. 336pp. 6 1/8 x 9 1/4.
0-486-47770-3

INTRODUCTION TO BIOSTATISTICS: Second Edition, Robert R. Sokal and F. James Rohlf. Suitable for undergraduates with a minimal background in mathematics, this introduction ranges from descriptive statistics to fundamental distributions and the testing of hypotheses. Includes numerous worked-out problems and examples. 1987 edition. 384pp. 6 1/8 x 9 1/4.
0-486-46961-1

Browse over 9,000 books at www.doverpublications.com

Mathematics–Geometry and Topology

PROBLEMS AND SOLUTIONS IN EUCLIDEAN GEOMETRY, M. N. Aref and William Wernick. Based on classical principles, this book is intended for a second course in Euclidean geometry and can be used as a refresher. More than 200 problems include hints and solutions. 1968 edition. 272pp. 5 3/8 x 8 1/2.   0-486-47720-7

TOPOLOGY OF 3-MANIFOLDS AND RELATED TOPICS, Edited by M. K. Fort, Jr. With a New Introduction by Daniel Silver. Summaries and full reports from a 1961 conference discuss decompositions and subsets of 3-space; n-manifolds; knot theory; the Poincaré conjecture; and periodic maps and isotopies. Familiarity with algebraic topology required. 1962 edition. 272pp. 6 1/8 x 9 1/4.   0-486-47753-3

POINT SET TOPOLOGY, Steven A. Gaal. Suitable for a complete course in topology, this text also functions as a self-contained treatment for independent study. Additional enrichment materials make it equally valuable as a reference. 1964 edition. 336pp. 5 3/8 x 8 1/2.   0-486-47222-1

INVITATION TO GEOMETRY, Z. A. Melzak. Intended for students of many different backgrounds with only a modest knowledge of mathematics, this text features self-contained chapters that can be adapted to several types of geometry courses. 1983 edition. 240pp. 5 3/8 x 8 1/2.   0-486-46626-4

TOPOLOGY AND GEOMETRY FOR PHYSICISTS, Charles Nash and Siddhartha Sen. Written by physicists for physics students, this text assumes no detailed background in topology or geometry. Topics include differential forms, homotopy, homology, cohomology, fiber bundles, connection and covariant derivatives, and Morse theory. 1983 edition. 320pp. 5 3/8 x 8 1/2.   0-486-47852-1

BEYOND GEOMETRY: Classic Papers from Riemann to Einstein, Edited with an Introduction and Notes by Peter Pesic. This is the only English-language collection of these 8 accessible essays. They trace seminal ideas about the foundations of geometry that led to Einstein's general theory of relativity. 224pp. 6 1/8 x 9 1/4.   0-486-45350-2

GEOMETRY FROM EUCLID TO KNOTS, Saul Stahl. This text provides a historical perspective on plane geometry and covers non-neutral Euclidean geometry, circles and regular polygons, projective geometry, symmetries, inversions, informal topology, and more. Includes 1,000 practice problems. Solutions available. 2003 edition. 480pp. 6 1/8 x 9 1/4.   0-486-47459-3

TOPOLOGICAL VECTOR SPACES, DISTRIBUTIONS AND KERNELS, François Trèves. Extending beyond the boundaries of Hilbert and Banach space theory, this text focuses on key aspects of functional analysis, particularly in regard to solving partial differential equations. 1967 edition. 592pp. 5 3/8 x 8 1/2.
0-486-45352-9

INTRODUCTION TO PROJECTIVE GEOMETRY, C. R. Wylie, Jr. This introductory volume offers strong reinforcement for its teachings, with detailed examples and numerous theorems, proofs, and exercises, plus complete answers to all odd-numbered end-of-chapter problems. 1970 edition. 576pp. 6 1/8 x 9 1/4.   0-486-46895-X

FOUNDATIONS OF GEOMETRY, C. R. Wylie, Jr. Geared toward students preparing to teach high school mathematics, this text explores the principles of Euclidean and non-Euclidean geometry and covers both generalities and specifics of the axiomatic method. 1964 edition. 352pp. 6 x 9.   0-486-47214-0

Browse over 9,000 books at www.doverpublications.com

Mathematics–History

THE WORKS OF ARCHIMEDES, Archimedes. Translated by Sir Thomas Heath. Complete works of ancient geometer feature such topics as the famous problems of the ratio of the areas of a cylinder and an inscribed sphere; the properties of conoids, spheroids, and spirals; more. 326pp. 5 3/8 x 8 1/2. 0-486-42084-1

THE HISTORICAL ROOTS OF ELEMENTARY MATHEMATICS, Lucas N. H. Bunt, Phillip S. Jones, and Jack D. Bedient. Exciting, hands-on approach to understanding fundamental underpinnings of modern arithmetic, algebra, geometry and number systems examines their origins in early Egyptian, Babylonian, and Greek sources. 336pp. 5 3/8 x 8 1/2. 0-486-25563-8

THE THIRTEEN BOOKS OF EUCLID'S ELEMENTS, Euclid. Contains complete English text of all 13 books of the Elements plus critical apparatus analyzing each definition, postulate, and proposition in great detail. Covers textual and linguistic matters; mathematical analyses of Euclid's ideas; classical, medieval, Renaissance and modern commentators; refutations, supports, extrapolations, reinterpretations and historical notes. 995 figures. Total of 1,425pp. All books 5 3/8 x 8 1/2.
Vol. I: 443pp. 0-486-60088-2
Vol. II: 464pp. 0-486-60089-0
Vol. III: 546pp. 0-486-60090-4

A HISTORY OF GREEK MATHEMATICS, Sir Thomas Heath. This authoritative two-volume set that covers the essentials of mathematics and features every landmark innovation and every important figure, including Euclid, Apollonius, and others. 5 3/8 x 8 1/2.
Vol. I: 461pp. 0-486-24073-8
Vol. II: 597pp. 0-486-24074-6

A MANUAL OF GREEK MATHEMATICS, Sir Thomas L. Heath. This concise but thorough history encompasses the enduring contributions of the ancient Greek mathematicians whose works form the basis of most modern mathematics. Discusses Pythagorean arithmetic, Plato, Euclid, more. 1931 edition. 576pp. 5 3/8 x 8 1/2.
0-486-43231-9

CHINESE MATHEMATICS IN THE THIRTEENTH CENTURY, Ulrich Libbrecht. An exploration of the 13th-century mathematician Ch'in, this fascinating book combines what is known of the mathematician's life with a history of his only extant work, the Shu-shu chiu-chang. 1973 edition. 592pp. 5 3/8 x 8 1/2.
0-486-44619-0

PHILOSOPHY OF MATHEMATICS AND DEDUCTIVE STRUCTURE IN EUCLID'S ELEMENTS, Ian Mueller. This text provides an understanding of the classical Greek conception of mathematics as expressed in Euclid's Elements. It focuses on philosophical, foundational, and logical questions and features helpful appendixes. 400pp. 6 1/2 x 9 1/4. 0-486-45300-6

BEYOND GEOMETRY: Classic Papers from Riemann to Einstein, Edited with an Introduction and Notes by Peter Pesic. This is the only English-language collection of these 8 accessible essays. They trace seminal ideas about the foundations of geometry that led to Einstein's general theory of relativity. 224pp. 6 1/8 x 9 1/4. 0-486-45350-2

HISTORY OF MATHEMATICS, David E. Smith. Two-volume history – from Egyptian papyri and medieval maps to modern graphs and diagrams. Non-technical chronological survey with thousands of biographical notes, critical evaluations, and contemporary opinions on over 1,100 mathematicians. 5 3/8 x 8 1/2.
Vol. I: 618pp. 0-486-20429-4
Vol. II: 736pp. 0-486-20430-8

Physics

THEORETICAL NUCLEAR PHYSICS, John M. Blatt and Victor F. Weisskopf. An uncommonly clear and cogent investigation and correlation of key aspects of theoretical nuclear physics by leading experts: the nucleus, nuclear forces, nuclear spectroscopy, two-, three- and four-body problems, nuclear reactions, beta-decay and nuclear shell structure. 896pp. 5 3/8 x 8 1/2. 0-486-66827-4

QUANTUM THEORY, David Bohm. This advanced undergraduate-level text presents the quantum theory in terms of qualitative and imaginative concepts, followed by specific applications worked out in mathematical detail. 655pp. 5 3/8 x 8 1/2.
0-486-65969-0

ATOMIC PHYSICS AND HUMAN KNOWLEDGE, Niels Bohr. Articles and speeches by the Nobel Prize–winning physicist, dating from 1934 to 1958, offer philosophical explorations of the relevance of atomic physics to many areas of human endeavor. 1961 edition. 112pp. 5 3/8 x 8 1/2. 0-486-47928-5

COSMOLOGY, Hermann Bondi. A co-developer of the steady-state theory explores his conception of the expanding universe. This historic book was among the first to present cosmology as a separate branch of physics. 1961 edition. 192pp. 5 3/8 x 8 1/2.
0-486-47483-6

LECTURES ON QUANTUM MECHANICS, Paul A. M. Dirac. Four concise, brilliant lectures on mathematical methods in quantum mechanics from Nobel Prize-winning quantum pioneer build on idea of visualizing quantum theory through the use of classical mechanics. 96pp. 5 3/8 x 8 1/2. 0-486-41713-1

THE PRINCIPLE OF RELATIVITY, Albert Einstein and Frances A. Davis. Eleven papers that forged the general and special theories of relativity include seven papers by Einstein, two by Lorentz, and one each by Minkowski and Weyl. 1923 edition. 240pp. 5 3/8 x 8 1/2. 0-486-60081-5

PHYSICS OF WAVES, William C. Elmore and Mark A. Heald. Ideal as a classroom text or for individual study, this unique one-volume overview of classical wave theory covers wave phenomena of acoustics, optics, electromagnetic radiations, and more. 477pp. 5 3/8 x 8 1/2. 0-486-64926-1

THERMODYNAMICS, Enrico Fermi. In this classic of modern science, the Nobel Laureate presents a clear treatment of systems, the First and Second Laws of Thermodynamics, entropy, thermodynamic potentials, and much more. Calculus required. 160pp. 5 3/8 x 8 1/2. 0-486-60361-X

QUANTUM THEORY OF MANY-PARTICLE SYSTEMS, Alexander L. Fetter and John Dirk Walecka. Self-contained treatment of nonrelativistic many-particle systems discusses both formalism and applications in terms of ground-state (zero-temperature) formalism, finite-temperature formalism, canonical transformations, and applications to physical systems. 1971 edition. 640pp. 5 3/8 x 8 1/2. 0-486-42827-3

QUANTUM MECHANICS AND PATH INTEGRALS: Emended Edition, Richard P. Feynman and Albert R. Hibbs. Emended by Daniel F. Styer. The Nobel Prize–winning physicist presents unique insights into his theory and its applications. Feynman starts with fundamentals and advances to the perturbation method, quantum electrodynamics, and statistical mechanics. 1965 edition, emended in 2005. 384pp. 6 1/8 x 9 1/4. 0-486-47722-3

Physics

INTRODUCTION TO MODERN OPTICS, Grant R. Fowles. A complete basic undergraduate course in modern optics for students in physics, technology, and engineering. The first half deals with classical physical optics; the second, quantum nature of light. Solutions. 336pp. 5 3/8 x 8 1/2. 0-486-65957-7

THE QUANTUM THEORY OF RADIATION: Third Edition, W. Heitler. The first comprehensive treatment of quantum physics in any language, this classic introduction to basic theory remains highly recommended and widely used, both as a text and as a reference. 1954 edition. 464pp. 5 3/8 x 8 1/2. 0-486-64558-4

QUANTUM FIELD THEORY, Claude Itzykson and Jean-Bernard Zuber. This comprehensive text begins with the standard quantization of electrodynamics and perturbative renormalization, advancing to functional methods, relativistic bound states, broken symmetries, nonabelian gauge fields, and asymptotic behavior. 1980 edition. 752pp. 6 1/2 x 9 1/4. 0-486-44568-2

FOUNDATIONS OF POTENTIAL THERY, Oliver D. Kellogg. Introduction to fundamentals of potential functions covers the force of gravity, fields of force, potentials, harmonic functions, electric images and Green's function, sequences of harmonic functions, fundamental existence theorems, and much more. 400pp. 5 3/8 x 8 1/2.
0-486-60144-7

FUNDAMENTALS OF MATHEMATICAL PHYSICS, Edgar A. Kraut. Indispensable for students of modern physics, this text provides the necessary background in mathematics to study the concepts of electromagnetic theory and quantum mechanics. 1967 edition. 480pp. 6 1/2 x 9 1/4. 0-486-45809-1

GEOMETRY AND LIGHT: The Science of Invisibility, Ulf Leonhardt and Thomas Philbin. Suitable for advanced undergraduate and graduate students of engineering, physics, and mathematics and scientific researchers of all types, this is the first authoritative text on invisibility and the science behind it. More than 100 full-color illustrations, plus exercises with solutions. 2010 edition. 288pp. 7 x 9 1/4. 0-486-47693-6

QUANTUM MECHANICS: New Approaches to Selected Topics, Harry J. Lipkin. Acclaimed as "excellent" (*Nature*) and "very original and refreshing" (*Physics Today*), these studies examine the Mössbauer effect, many-body quantum mechanics, scattering theory, Feynman diagrams, and relativistic quantum mechanics. 1973 edition. 480pp. 5 3/8 x 8 1/2. 0-486-45893-8

THEORY OF HEAT, James Clerk Maxwell. This classic sets forth the fundamentals of thermodynamics and kinetic theory simply enough to be understood by beginners, yet with enough subtlety to appeal to more advanced readers, too. 352pp. 5 3/8 x 8 1/2. 0-486-41735-2

QUANTUM MECHANICS, Albert Messiah. Subjects include formalism and its interpretation, analysis of simple systems, symmetries and invariance, methods of approximation, elements of relativistic quantum mechanics, much more. "Strongly recommended." – *American Journal of Physics.* 1152pp. 5 3/8 x 8 1/2. 0-486-40924-4

RELATIVISTIC QUANTUM FIELDS, Charles Nash. This graduate-level text contains techniques for performing calculations in quantum field theory. It focuses chiefly on the dimensional method and the renormalization group methods. Additional topics include functional integration and differentiation. 1978 edition. 240pp. 5 3/8 x 8 1/2.
0-486-47752-5

Browse over 9,000 books at www.doverpublications.com

Physics

MATHEMATICAL TOOLS FOR PHYSICS, James Nearing. Encouraging students' development of intuition, this original work begins with a review of basic mathematics and advances to infinite series, complex algebra, differential equations, Fourier series, and more. 2010 edition. 496pp. 6 1/8 x 9 1/4. 0-486-48212-X

TREATISE ON THERMODYNAMICS, Max Planck. Great classic, still one of the best introductions to thermodynamics. Fundamentals, first and second principles of thermodynamics, applications to special states of equilibrium, more. Numerous worked examples. 1917 edition. 297pp. 5 3/8 x 8. 0-486-66371-X

AN INTRODUCTION TO RELATIVISTIC QUANTUM FIELD THEORY, Silvan S. Schweber. Complete, systematic, and self-contained, this text introduces modern quantum field theory. "Combines thorough knowledge with a high degree of didactic ability and a delightful style." – *Mathematical Reviews.* 1961 edition. 928pp. 5 3/8 x 8 1/2. 0-486-44228-4

THE ELECTROMAGNETIC FIELD, Albert Shadowitz. Comprehensive undergraduate text covers basics of electric and magnetic fields, building up to electromagnetic theory. Related topics include relativity theory. Over 900 problems, some with solutions. 1975 edition. 768pp. 5 5/8 x 8 1/4. 0-486-65660-8

THE PRINCIPLES OF STATISTICAL MECHANICS, Richard C. Tolman. Definitive treatise offers a concise exposition of classical statistical mechanics and a thorough elucidation of quantum statistical mechanics, plus applications of statistical mechanics to thermodynamic behavior. 1930 edition. 704pp. 5 5/8 x 8 1/4. 0-486-63896-0

INTRODUCTION TO THE PHYSICS OF FLUIDS AND SOLIDS, James S. Trefil. This interesting, informative survey by a well-known science author ranges from classical physics and geophysical topics, from the rings of Saturn and the rotation of the galaxy to underground nuclear tests. 1975 edition. 320pp. 5 3/8 x 8 1/2. 0-486-47437-2

STATISTICAL PHYSICS, Gregory H. Wannier. Classic text combines thermodynamics, statistical mechanics, and kinetic theory in one unified presentation. Topics include equilibrium statistics of special systems, kinetic theory, transport coefficients, and fluctuations. Problems with solutions. 1966 edition. 532pp. 5 3/8 x 8 1/2. 0-486-65401-X

SPACE, TIME, MATTER, Hermann Weyl. Excellent introduction probes deeply into Euclidean space, Riemann's space, Einstein's general relativity, gravitational waves and energy, and laws of conservation. "A classic of physics." – *British Journal for Philosophy and Science.* 330pp. 5 3/8 x 8 1/2. 0-486-60267-2

RANDOM VIBRATIONS: Theory and Practice, Paul H. Wirsching, Thomas L. Paez and Keith Ortiz. Comprehensive text and reference covers topics in probability, statistics, and random processes, plus methods for analyzing and controlling random vibrations. Suitable for graduate students and mechanical, structural, and aerospace engineers. 1995 edition. 464pp. 5 3/8 x 8 1/2. 0-486-45015-5

PHYSICS OF SHOCK WAVES AND HIGH-TEMPERATURE HYDRO DYNAMIC PHENOMENA, Ya B. Zel'dovich and Yu P. Raizer. Physical, chemical processes in gases at high temperatures are focus of outstanding text, which combines material from gas dynamics, shock-wave theory, thermodynamics and statistical physics, other fields. 284 illustrations. 1966–1967 edition. 944pp. 6 1/8 x 9 1/4. 0-486-42002-7

Browse over 9,000 books at www.doverpublications.com

CPSIA information can be obtained
at www.ICGtesting.com
Printed in the USA
LVOW13s1307120318
569542LV00019B/501/P